472-5872

Selected Topics In Mathematics

Edward L. Spitznagel, Jr.
Washington University

HOLT, RINEHART AND WINSTON, INC.
*New York   Chicago   San Francisco   Atlanta
Dallas   Montreal   Toronto   London   Sydney*

Copyright © 1971 by Holt, Rinehart and Winston, Inc.
All rights reserved
Library of Congress Catalog Card Number: 74-141514
AMS 1970 Subject Classifications 00-01, 00A05, 05-01, 10-01, 15-01, 26A06,
55-01, 60-01, 62-01, 68-01
ISBN 0-03-084693-5
Printed in the United States of America
6 038 987654

Preface

This book is designed for use in either a half or a full year mathematics survey course at the beginning college level. It is based on a course that I taught to approximately 160 students at Northwestern University during the academic year 1968–1969.

I became interested in developing the approach in this book after seeing what little interest students (justifiably) show in logic and sets and integral domain axioms and algebra-trigonometry rehashes and all the rest of the topics which are standard fare in such courses. My solution was to teach only topics that the students could see the good of within the first few days. And, knowing that every bit of motivation helps, I always included human interest items and historical references where they were appropriate.

The results were most encouraging. I had been apprehensive of how members of the class whose CEEB mathematics scores ranged from 509 to 635 would do, but I found they had no particular difficulty. These students were the ones to whom I particularly addressed my course, and I have written this book at the level I believe most suited to them. The material presupposes only high school algebra (no quadratic formula), though, of course, some high school geometry is helpful. The chapters on probability and statistics are written in such a fashion that no square roots need be taken.

Since liberal arts mathematics courses in different schools vary a great deal in length, this book is designed so that one half of the material can be lifted out to make up a one quarter or one semester course. The organization of the book is displayed in the following table:

	Topic	Chapters
	What is Mathematics?	1
Major	Number Theory	2, 3
	Topology	4, 5
	Recreations	6, 7
	Matrices	8, 9
	Probability and Statistics	10, 11
Minor	Computer Programming	12
	Introduction to Calculus	13

In each major topic, the second chapter has the first as a prerequisite. Other than that, there is no interdependence of chapters. Thus, the first (easier) chapter of each major topic can be lifted out and put together with Chapter 1 to form a balanced short course, consisting of Chapters 1, 2, 4, 6, 8, and 10. There are a number of other good short courses that one can teach from the book, depending on student interest.

Judging from my own experience, the material "goes" naturally at about one section every two class hours. That includes the time spent discussing selected exercises. Speed will of course vary with each group of students as well as with the topics covered.

Special comments are in order concerning Chapter 11 (statistics) and Chapter 12 (programming). The statistical tests presented in Chapter 11 are of the "nonparametric" variety. They are easy to teach and are actually enjoyable to use—not at all like the more familiar parametric tests. They are nearly as powerful as the parametric tests, and their use is justified in many situations where the parametric tests are not reliable. I believe this is the first book at the beginning college level to present such tests. However, I know of colleagues who have taught them with great success even to high school classes. If time permits, I highly recommend having each student do a research project utilizing one or more of the tests in Chapter 11. For most of my students, their research projects were the high point of the course; the quality of their work can be seen by looking at the Chapter 11 exercises, most of which were student research projects. A questionnaire given to the entire class is a convenient way to collect data. For most purposes, any sample size over 100 is adequate.

The chapter on computer programming introduces students to the basic processes of looping and branching, which are responsible for the power of computers. Since the only things needed to write simple loops are floating-point arithmetic and the arithmetic IF, I went no further into FORTRAN than these. The chapter can be used with both FORTRAN IV and the older FORTRAN II. If you let your students write and run programs (which I heartily recommend), it is a good idea to keep tight rein on the time limit, since all too often loops made with the IF statement at first contain mistakes which make them endless.

In my course, whenever I could ask the students to do something purposeful and creative (and not too time-consuming), I did so. Also, whenever I could do something in class besides the usual writing on the blackboard, I took advantage of the opportunity. I braided the slit strip of paper according to the method described in Chapter 6. (I told them I could also demonstrate with a girl's hair, but there were no volunteers.) I did a short ESP experiment with + and − cards to illustrate a binomial hypothesis test. (The result was significant at the .05 level, to the embarrassment of skeptical me.)

With their air pucks, pendulums, explosions, and color changes, physicists and chemists are naturally one-up on mathematicians. However, with a little creativity on our part, there is no reason we cannot develop a creative series of demonstrations to stimulate our classes. I would appreciate if anyone using this book would tell me of demonstrations he found helpful in class. If enough of them are forthcoming, perhaps they could be collected into a "gimmick" manual to supplement the instructor's manual.

I wish to thank Northwestern University for granting me a leave of absence during which a portion of this book was written, New Mexico State University and the University of California at Santa Cruz for the use of their facilities during my leave, and the Washington University Computing Facilities for computer time (through N.S.F. Grant G-22296) used in writing this book and in preparing its tables.

Professors Joseph Dorsett, Victor Klee, and Jack Robertson read a first draft of the entire book. Their suggestions for improvement were most helpful, and their comments were very encouraging. I am also indebted to the staff of Holt, Rinehart and Winston for all their help from beginning to end of the publication process.

St. Louis, Missouri   Edward L. Spitznagel, Jr.
January, 1971

Contents

Preface v

1 What Is Mathematics? 1

1. Introduction 1
2. Mathematics 2
3. Theorems and Proofs 6
4. The Philosophy behind Exercises 12

2 First Steps in Number Theory 15

1. Introduction 15
2. Primes, Composites, and the Infinitude of the Composites 17
3. The Infinitude of the Primes 22
4. Prime-Generating Formulas 24
5. The World's Largest Prime 29
6. Euclid's Algorithm 32

3 Pythagorean Triples 42

1. The Theorem of Pythagoras 42
2. Formulas for Constructing Pythagorean Triples 48
3. Preliminaries 51
4. Proof 55
5. Fermat's Last Problem 58

4 Pioneering Results in Topology   60

1. Introduction   60
2. The Königsberg Bridge Problem   63
3. Duality   67
4. An Application of Graphs to Business Management   70
5. The Descartes-Euler Formula for Polyhedra   78
6. Regular Polyhedra   85

5 Map Coloring   92

1. Introduction   92
2. Some Properties of a Map when Viewed as a Graph   95
3. Application of the Descartes-Euler Formula   98
4. Reduction to the Extra Ordinary Case   101
5. Proof of the 5-Color Theorem for Ordinary Maps   102
6. Maps with Ring-Shaped Countries, Maps on the Surface of a Sphere, Maps on the Surface of a Doughnut   106

6 Miscellaneous Topics   111

1. Introduction   111
2. Fiftieth Anniversaries   111
3. Knots, Braids, and a Surprise   114
4. Matching Birthdays   118
5. Casting Out Nines   122

7 Topics Having To Do with the Number Two   128

1. Binary Notation   128
2. Nim   132
3. The Tower of Brahma   137
4. The Fifteen Puzzle   142
5. Establishing the Difference between Even and Odd Rearrangements   150

8 Matrices   152

1. Introduction to Matrices and Their Arithmetic   152
2. The Adjacency Matrix of a Graph   158
3. Predicting Populations   164
4. Markov Chains   171

9 Row Reduction of Matrices   179

1. Introduction   179
2. Row Reduction by Example   181
3. Markov Chains Revisited   188

CONTENTS xi

10 Introduction to Probability Theory 193

1. Introduction 193
2. Flipping a Fair Coin 195
3. A Central Limit Theorem 202
4. Introduction to Hypothesis Testing 205
5. The Notion of a Test Statistic 212

11 Three Statistical Tests 215

1. Introduction 215
2. The Role of Statistical Tests in Research 216
3. The Pearson Test: Testing for a Relation between Two Conditions 217
4. The Wilcoxon Test: Testing for a Relation between a Condition and a Rating 225
5. The Kendall Test: Testing for a Relation between Two Ratings 235
6. Table A and Sample Sizes 243

12 Computer Programming 247

1. Introduction 247
2. FORTRAN Formulas 248
3. Assignment Statements and Our First Program 252
4. Branching and Looping 256
5. An Example of a Nested Loop. Flowcharting 259
6. Another Example of a Nested Loop 266
7. Computers — Past, Present, and Future 271
8. Card Programming 274

13 An Introduction to Calculus 277

1. Introduction 277
2. Numbers, Functions, and Graphs 278
3. Rate of Change 285
4. The Slope of a Tangent Line. Maximum and Minimum 292
5. Sine and Cosine 299
6. The Foundations of Calculus 305

Appendix 309

Table A 310
Table B 312
Table C 314

Index 315

Selected Topics In Mathematics

CHAPTER ONE

What Is Mathematics?

1. Introduction

For centuries men had dreamed of transmuting the elements, of being able to change "base" substances into pure gold. So strongly did they believe in the dream that the seekers, the alchemists, had already given a name to the means they hoped to find. They called it the Philosopher's Stone.

In our own century the dream of the alchemists finally came true. The first machine capable of transmuting the elements was completed and used in the year 1932 by two young English physicists, John Cockcroft and E. T. S. Walton. Their success, like most of men's greatest accomplishments, was not the result of haphazard experimentation. Rather, men had been working toward the goal for three centuries by slowly accumulating knowledge in the sciences which made possible the machine of Cockcroft and Walton.

The ultimate means to all of man's greatest achievements, the real Philosopher's Stone, is his capacity to reason. Reason enables us to go beyond our observations of the physical world, to probe beneath what we see and discover orderly patterns in the behavior of our world. The products of man's reasoning abilities have been classified into a number of subjects called *sciences*. The classification is really after the fact; there are no sharp dividing lines, but simply rough boundaries between the areas in which men's reasoning powers have been used in the past. This book is about one of those sciences, one area in which men have successfully used their intellectual powers, one facet of the Philosopher's Stone—mathematics.

2. Mathematics

As the term is presently defined, mathematics is that science which is concerned primarily with quantitative, spatial, and logical relationships. It deals with these relationships in a very abstract setting, making use of symbols and special terms to represent the ideas under study. Because of its abstract, unprejudiced view of quantity and space, mathematics has been found to be applicable to any other study in which quantitative or spatial relations are of importance. Since one or both kinds of these relations are studied in nearly every science, mathematics has become a tool common to all of the sciences.

Of course, the generality or abstraction that makes mathematics so useful *can* also make it difficult to study. We human beings seldom think purely in terms of symbols and words; we also borrow sense images or pictures from our memories in the process of our thinking. In order to take advantage of his full thought apparatus, a mathematician frequently constructs concrete examples of whatever he is studying. These examples might be numbers arranged on paper, figures drawn on paper, or similar things visualized in his mind. Frequently, a mathematician's creativity is directly related to his ability to think up and work with examples. We might describe the trick of using examples as an adaptation of mathematics to human nature.

On the other hand, it is also necessary to adapt human nature to the demands of mathematics. We are all accustomed to the practice of reasoning according to certain rules of logic—*more or less*. In day-to-day life, we may now and again disobey a rule of logic, yet be guided to a correct conclusion by our past experiences. But mathematics is sufficiently abstracted from day-to-day living that past experience is no safe guide through its intricacies. In fact, if a person depends upon anything else but rules of logic in mathematics, he is almost certain of reaching false conclusions.

Thus, while one *thinks* about mathematics in terms of examples and past experiences, he is very careful to use nothing but rules of logic in drawing conclusions. This care has led to the writing of much mathematical reasoning in the form of theorems followed by proofs.

2. MATHEMATICS

A theorem is a statement of a conclusion that a mathematician has reached. The proof which he writes down below the theorem is what he believes to be a logically perfect chain of reasoning leading to his conclusion.

In keeping with the demand for logical perfection, the mathematician also admits that he cannot prove a theorem without reasoning to it from some other assertions. Somewhere, at the very beginning of his subject, there must be statements which he simply accepts as true, in order that he may prove theorems from them. These statements, called *axioms*, are always chosen to be as simple as possible and as few in number as possible. Usually, the axioms are chosen only after the subject has been studied for some time.

We have mentioned how dependent mathematics is upon reason. In the various ages of recorded history, mathematical creativity has had its ups and downs, depending upon how free men were to follow pure reason. First of all, one needed to have some hours free from the effort and worry of obtaining his next meal. Second, one needed a mind free enough to challenge existing belief, if his reason said such belief was wrong.

The three outstanding periods and places of mathematical advance that are known to us are: (1) 2000 to 1600 B.C. in Mesopotamia, (2) 600 B.C. to 300 A.D. in and around Greece, and (3) 1600 A.D. onward beginning in western Europe and spreading to the rest of the globe.

About the first period, we wish we knew much more than is presently known. All we have are some clay tablets containing some truly remarkable series of numbers. The inhabitants of Mesopotamia did mathematical calculations in a base sixty number system, as opposed to our base ten system. The 60 minutes to an hour, the 60 seconds to a minute, and the 360 degrees to a circle are all traceable back to them.

You know about the second period from your history courses. The Greeks had a view of living rather different from ours. Most of their creative work—in literature, philosophy, mathematics, and so on—was done not for livelihood, but for its own sake, as leisure activity. The Greeks worked at making a living only enough to ensure reasonable comfort. Any attempt by a person to amass wealth for its own sake would have been viewed as positively indecent.

The third period began with a storm over a new theory of the earth's position in the universe. The accepted theory was that the earth was the center of everything, and that theory fit in very nicely with the view of man as the center of all creation. The fact that the earth-centered theory was attacked and at last overthrown was in part a sign that men were once more beginning to trust their own powers of reason and observation over the opinions of authority. It was also a direct cause of the rebirth of mathematical creativity, because the new view of the solar system led to Newton's discovery of calculus as the mathematical tool to be used in the study of the motions of the planets. From the discovery of calculus to the present day, the intensity of mathematical creativity has been on the increase.

This book contains mathematics originating from the three periods of creativity we have mentioned. That it does is somewhat accidental, because the book is designed not as a historical presentation of mathematics but rather as a collection of some outstanding ideas in mathematics. The ultimate choice of topics was made on the basis of opinions of students like you, to whom I taught an experimental course from which this book was developed.

Their primary concern was to be taught the kinds of mathematics in which they could readily see the good of the subject. Some topics that had great appeal were quite "pure"—of interest mainly because of their beauty. Others were quite applied, down-to-earth. I was happy to be able to make this book balanced both over the historical periods and over the pure-to-applied spectrum, yet make it up entirely from topics that had appealed to my students.

In each of the topics covered in this book, you will have an opportunity to do some of the mathematics yourself—as early as the first day or two days into the topic. In the chapter on statistics you will learn three very useful hypothesis tests, which you will be able to apply to real data. The exercises contain a number of interesting sets of data to which you can apply the tests, but you can also apply them to data from your psychology or sociology courses. In the chapter on computer programming, you will learn how computers function by actually writing simple but practical FORTRAN programs. Those programs will run—exactly as they are written—on almost any digital computer in existence. In the chapters on number theory and topology, you will have the opportunity of making up examples and writing down your own proofs. That is, you will have the opportunity to do the same kind of thinking that the pioneers in these fields did.

Many of the exercises you will be doing are unlike the kinds of drill manipulations commonly taught in grade school and high school mathematics courses. At times you may feel a bit uneasy when you are asked to *discover* how to work a problem, instead of being told to work ten drill exercises by the method on page x. But the difference between imitation and discovery is really the difference between computation and mathematics. In grade school and high school you were primarily taught those computational skills that are necessary for survival in our society. In this course, you will learn some more of those methods, for they are important, but you will also begin to learn the more creative skills that are the real basis of mathematics.

SUGGESTED READINGS

Bell, E. T. *The Last Problem.* New York: Simon and Schuster, 1961. See especially pages 22–39, 44–101, and 274–296.
   A very lively history of some periods of mathematical creativity.

―――― *Men of Mathematics.* New York: Simon and Schuster, 1937.
   The best-written collection of mathematical biographies in the English language—perhaps in any language.

de Santillana, G. *The Crime of Galileo.* Chicago: The University of Chicago Press, 1955.
   Galileo has been called the father of modern science. His arguments in favor of the sun as the center of the planetary system caused him to be tried for heresy. He recanted and then was sentenced to spend the remainder of his life under house arrest. The preface to the book gives an outline of the complexities of the trial.

Halmos, P. R. "Nicolas Bourbaki," in *Mathematics in the Modern World,* ed. by M. Kline. San Francisco: W. H. Freeman Co., 1968, pp. 77–81.
   The biography of a famous but nonexistent mathematician.

Keynes, J. M. "Newton, the Man," in *The World of Mathematics,* ed. by J. R. Newman. New York: Simon and Schuster, 1956, pp. 277–285.
   An unusual and startling portrait of England's greatest mathematician.

Neugebauer, O. *The Exact Sciences in Antiquity.* Providence: Brown University Press, 1957, pp. 53–66.
   This third chapter of the book describes the kinds of written records of ancient mathematics that have survived as much as forty centuries of time. It also describes present-day circumstances that make life difficult for those who try to study these records.

Newman, J. R. "Srinivasa Ramanujan," in *Mathematics in the Modern World,* ed. by M. Kline. San Francisco: W. H. Freeman Co., 1968, pp. 73–76.
   Portrait of a self-taught mathematician.

3. Theorems and Proofs

The high level of abstraction in mathematics makes it necessary that mathematical reasoning be especially airtight, that each new fact be derived from previous ones solely by rules of logic. In mathematical subjects where the build-up of fact upon fact becomes especially complex, an organization of the subject matter into major facts, called *theorems*, each followed by its logical derivation, or *proof*, generally helps a person to see what is going on.

The theorem-proof format can be a source of irritation to a reader, unless he knows that it is serving a legitimate purpose. All mathematical subjects, by their very nature, consist of logical derivations, or proofs. But some, usually the ones called "pure" subjects, must rely on the theorem-proof structure more than others. In these it may take a long, somewhat zigzagging chain of reasoning to reach a particular conclusion. If this is the case, the mathematical writer will state the conclusion first, calling it a theorem, and then he will give the proof of his statement. In mathematics, just as in day-to-day life, seeing one's goal can be a definite help in achieving it.

Of course, the use of theorems and proofs does not make mathematical reasoning instantly and perfectly clear. It may be possible to follow the logic of a proof; however, the fact that someone can agree with each step does not guarantee that he will grasp the idea of the proof, nor that he will be able to remember it after he finishes reading it. The reason is that we are not completely cold, calculating beings. Our minds can work in a purely logical fashion when we make the effort, but usually they work by using a mixture of logic and memories of examples. In order to understand mathematical reasoning, and in order to remember the steps in mathematical reasoning, it is perfectly natural to "push" a concrete example through the reasoning process (that is, to apply each step of the general proof to a specific case) and see what happens.

Also, if we happen to be hunting for a proof of something we believe to be true, the study of an example or two may reveal the key idea behind the proof we are looking for. The validity of a proof will never depend in any way upon an example. By *its* very nature, a proof must consist of a logical derivation that anyone will agree with step by step. But by *our* very human nature, we will see the proof more clearly if we use the full capacity of our minds to understand it.

I place special emphasis on the use of examples because I have often noticed students failing to take full advantage of their powers of memory and visualization. You might like to try an experiment to confirm or reject this observation in your own case. Study a sample

3. THEOREMS AND PROOFS

of your own writing, other than a narrative, to see how many abstract words like situation, relevance, power, truth, love, and fear you find. Make special note of any abstract term that you use more than a few times. Then do the same with one or more authors in whom you are currently interested. If you find there is a difference, it will most likely be that your favorite authors keep a balance between abstract and concrete (image) words, while your own sample makes a heavier use of abstract words.

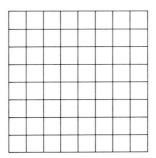

FIGURE 1. An 8 by 8 checkerboard (uncolored).

Let us now study an example of a theorem and its proof. Suppose we have a square checkerboard with an even number of squares along each side (as, for instance, the standard 8 by 8 checkerboard in Figure 1). Given a sufficient number of dominoes each the size of two checkerboard squares, it is possible to cover the entire board with the dominoes, each domino covering two squares, and each square covered exactly once. (For example, this can be done by covering each row individually.) Now, if we were to cut two diagonally opposite corner squares off the board as in Figure 2, could we still cover the board with dominoes?

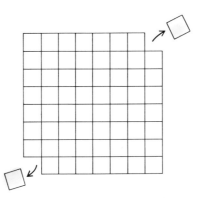

FIGURE 2. An 8 by 8 checkerboard with two diagonally opposite corner squares removed.

8   CHAP. 1 WHAT IS MATHEMATICS?

THEOREM. If two diagonally opposite corner squares of the checkerboard are clipped off, the remainder of the board cannot be covered with dominoes in the manner described above.

Proof. First color the squares of the full checkerboard alternately black and white, in the normal checkerboard pattern. Exactly half of the squares on the board are white, and exactly half are black. The pairs of diagonally opposite corner squares are of the same color, both squares white in the one pair, and both squares black in the other. Thus, when two diagonally opposite corner squares are removed as in Figure 3, the remainder of the board will have two more squares of one color than it will of the other.

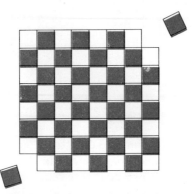

FIGURE 3. An 8 by 8 colored checkerboard with two diagonally opposite corner squares removed.

Now suppose it has been possible to cover the corner-clipped board with dominoes, each domino covering two squares, and each square covered exactly once. Each domino covers two adjacent squares, and on a checkerboard any two adjacent squares are of *different* colors. Therefore each domino covers exactly one white square, so the corner-clipped board must have as many white squares as there are dominoes covering the board. By the same argument, the board must have exactly as many black squares as there are dominoes covering it. Therefore, the board must have exactly as many black squares as it has white squares. This conclusion contradicts what we proved in the paragraph above. Therefore, the assumption that led to this conclusion—the assumption that we could cover the corner-clipped board—must be false. This completes the proof of the theorem.

Now, was the proof easy or hard to follow? However easy or hard it was for you, it probably would have been harder to follow had we not drawn a picture of a board with two of its corners removed. Yet the drawing of the picture does not justify any one of the steps of the proof. (In particular, the picture is not needed to show that diagonally opposite corner squares have the same color; we know that because all squares on any diagonal must have the same color.) The steps can be followed without any picture being drawn, because they are all logically correct.

Going a step further, do you think you could have invented this proof by yourself? You could have, very definitely, if you had simply been given the statement of the theorem and had spent half an hour or so trying it out on a 4 by 4 checkerboard.

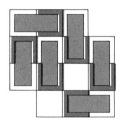 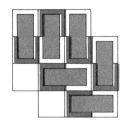 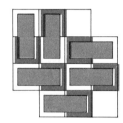

FIGURE 4. Three examples of failure to cover a corner-clipped 4 by 4 checkerboard with dominoes. In each case, the two squares remaining uncovered are both of the same color, the color opposite to the color of the two corners that were removed.

Suppose we do that right now. Take a 4 by 4 board and cut off two diagonally opposite corner squares, then try to cover the remainder with dominoes. After trying three different covering patterns, we always find ourselves frustrated. (See Figure 4.) But there is a pattern to the frustrations. Every single time, the squares left uncovered were both of the same color. Now why should that be? Taking a careful look at the squares we cut off, we discover that they are both of the same color, the color opposite to the one of the two squares we failed to cover. Then the explanation dawns on us. Each domino has to cover a square of each color. We have unbalanced the colors by removing two squares of the same color. Therefore, there will always be two squares, of the color opposite to that of the two we removed, that we will not be able to cover with a domino.

We did, of course, make use of the color pattern of a checkerboard in our rediscovery of the idea of the proof. Without that pattern, the rediscovery would have been much harder. The moral is: Think concretely. If the theorem talks about a checkerboard, think about a checkerboard, complete with pattern. Had we not bothered to think about the pattern from the beginning, we still could have found the proof (by finding that the pattern played a role), but the discovery would have taken longer. What would have happened, probably, is that we would have tried perhaps half a dozen coverings, usually ending with two uncovered squares touching each other only on a corner, so a domino would not cover them.

Then we would have asked why that should happen so often. A sufficiently long stare at the board would reveal that the corner squares removed were also in diagonal positions relative to each other, just like the two uncovered squares touching on a corner, and *then* we might begin to think of the diagonal coloring pattern of the checkerboard. But, of course, that process would take much longer than would the discovery of the proof if we were thinking more concretely in terms of the colored checkerboard.

EXERCISES

1. Suppose we cut two diagonally opposite corners off a 7 by 7 checkerboard. Can we cover the remainder of the board with dominoes with each domino covering two squares and each square covered exactly once? Why or why not? [*Hint:* This is much easier than the theorem we proved in the text.]

2. A "tromino" is a rectangle three times as long as it is wide. Show that a 9 by 9 checkerboard can be covered with trominoes, each tromino covering 3 squares and each square covered exactly once.

3. Show that if three of the four corner squares are clipped off the 9 by 9 checkerboard, the remainder of the checkerboard cannot be covered with trominoes in the manner prescribed in Exercise 2.

4. An angle tromino is made by placing three squares of the same size together in the shape of an L. Show that if three of the four corner squares are clipped off a 3 by 3 checkerboard, the remainder of the checkerboard can be covered by 2 angle trominoes.

5. Show that if three of the four corner squares are clipped off a 6 by 6 checkerboard, the remainder of the checkerboard can be covered by 11 angle trominoes.

The remaining exercises of this section are designed to lead you through a second example of a theorem and its proof. The square root of 2, written $\sqrt{2}$, is defined to be that positive number n such that $n^2 = n \cdot n = 2$. To ten decimal places, $\sqrt{2} = 1.4142135623 \cdots$. Back in the days when numbers were very much associated with mystery, mysticism, and magic, members of the Pythagorean Brotherhood discovered that $\sqrt{2}$ could not be represented by a fraction a/b, where a and b are whole numbers. Until that discovery, all mathematicians had believed that every number was *rational*, the ratio of two whole numbers.

One of the basic tenets of the Pythagorean doctrine was that whole numbers were the key to understanding the universe. The discovery that some numbers were not fractions of whole numbers made their entire philosophy look fishy. The Pythagoreans tried to keep the irrationality of $\sqrt{2}$ a secret within the brotherhood, but someone leaked it. (Legend has it that the man who let the secret out was drowned at sea for his offence.)

The proof that the square root of 2 is not rational is a nice example to give in a series of exercises. It uses only simple ideas from arithmetic, so the individual steps are easy to work out. The proof is a classic example of what is called the "indirect method." We will begin the proof by assuming that the statement we hope to prove is false. Then we will logically deduce from that assumption a statement that we *know* to be false. The fact that we will reach a false conclusion will mean that the assumption made at the beginning must be incorrect. Since that assumption is the opposite of what we are hoping to prove, it will follow that the statement we hope to prove is true.

THEOREM. The square root of 2 is irrational.

6. As we mentioned, we are going to show that assuming the opposite of this theorem leads to a conclusion we will know to to be false. The opposite of the theorem is that $\sqrt{2} = a/b$, where a and b are both whole numbers. Show first that if we have $\sqrt{2} = a/b$, where a and b are whole numbers, then also $\sqrt{2} = c/d$, where c and d are both whole numbers and at least one of c and d is an odd number.

7. Now, saying that $\sqrt{2} = c/d$ is the same as saying that $2 = c^2/d^2$, which is the same as saying that $2d^2 = c^2$. From this last equation prove that c must be an even number. [*Hint:* Use the indirect method, making use of the fact that the product of two odd numbers is again an odd number.]

8. Then prove that c^2 is a multiple of 4.
9. Then prove that d^2 is an even number.
10. Then prove that d must also be an even number.
11. State what you proved in Exercises 6, 7, and 10 as a single conclusion. Can this conclusion possibly be true? If it cannot, you have succeeded in giving an indirect proof of the theorem.

4. The Philosophy behind Exercises

Most exercises in this book fall into four main categories: Drill, Example, Proof, and Guess 'n' Proof.

Drill exercises are the sort of exercises everybody does and everybody hates. They are used to build necessary computational skills, but they do little else. They often give the false impression that all mathematics is strictly computational — which it certainly is not. Since a moderate amount of computational skill is all that is required for this book, and since you have already acquired most of that skill in grade and high school, you will not be seeing many exercises here that are merely drill.

Example exercises, like drill, are very down-to-earth work, but they serve a much better purpose. We have already discussed the value of examples in shedding light on mathematical reasoning. The example exercises in this book suggest specific examples which make the theory easier to grasp. You are, of course, free to make up your own examples as well, and I strongly encourage you to do so.

Proof exercises give you a chance to discover chains of mathematical reasoning on your own. What we said in the last section about studying an example to discover a proof can be applied to any exercise of this sort. However, there is another idea which can be used along with that on many of the proof exercises. Many proof exercises ask you to prove a statement which resembles very closely something proved in the text. The purpose of giving you such an exercise is to let you modify an existing proof in order to prove another result.

In doing such an exercise, you therefore already have a guide to follow — namely, the proof given in the text. By learning how you can alter it to reach some new conclusion, you will acquire additional insight into its workings. So when you see that you are asked to prove something similar to a result in the text, try making use of the proof you have already seen.

4. THE PHILOSOPHY BEHIND EXERCISES

Guess 'n' proof is the highest form of mathematical exercise. There are not very many such exercises in this book, because it takes just the right circumstances in the development of a subject for a student to guess that he will be able to prove such-and-such a result. When you do such an exercise, you will be participating in the creation of mathematics. No matter if someone has done it before; for you it will be original.

When we see mathematics today, with its efficient means of computing and its well-organized theories, we often lose sight of the processes that led to mathematics as it is now. Just about all of mathematics arose from people making guesses in various directions and patiently searching for proofs of these guesses. Those guesses which "panned out" are what make up the subject matter of mathematics today. The process of guessing and proving new results still goes on today, in fact at a speed far greater than in any previous century. All the mathematics presently known has opened a wide variety of new fields in which mathematicians are doing research. However, to do mathematical research today, one must have extensive knowledge of the mathematics which has been done in the past. Typically, a young man will study until age twenty-five or so before he has sufficient preparation to do research.

As well as the guess 'n' proof exercises, there are sometimes pure guess exercises. In these exercises, the process of finding a proof may be very long and difficult, but on the basis of examples you may be able to guess what could be proved. Never be afraid to make an educated guess. Even if your guess is wrong in the strict sense, there can still be a grain or a bushel of truth in it.

Lastly, use the above classifications only as broad guidelines to the kinds of exercises you will be doing. Most of the exercises in this book were made up before this section was written, so the descriptions are really after the fact. Most exercises do fit into the four categories fairly well, but some are hybrids, and some are simply different.

EXERCISES

1. (Sample of a drill exercise.) Calculate:

    1257^2 751^2 32^2 $11,111^2$.

2. (Sample of an example exercise.) Calculate $1^2, 2^2, 3^2, 4^2, \cdots, 18^2, 19^2, 20^2$. List the right-most digits (the "units" digits) appearing in each of the twenty numbers you have just calculated.

[*Note:* This exercise is an example with reference to the two exercises below.]
3. (Sample of a proof exercise.) Show that if a number has 3 as its right-most digit, then it cannot be the square of a whole number.
4. (Example of a guess 'n' proof exercise.) Guess which of the digits 0, 1, 2, 3, $\cdots$, 8, 9 cannot be the right-most digit in the square of a whole number. Prove your guess. [*Hint for Exercises 3 and 4:* Show that the right-most digit of the whole number n^2 is determined by knowledge of the right-most digit of n.]

CHAPTER TWO

First Steps in Number Theory

1. Introduction

In this chapter and the one to follow, we will discuss some intriguing properties of the so-called *natural numbers;* 1, 2, 3, 4, $\cdots$. It is especially fitting to begin a survey of mathematics with the theory of natural numbers, for this subject was one of the very first in which men gave proofs of certain assertions. That is, the theory of the natural numbers is one of the earliest examples of a mathematical theory, according to our understanding today of what is mathematics.

Until recently, the beginning of the theory of natural numbers was credited to the Greeks of the era immediately preceding that of their greatest philosophers. However, it now appears that the Babylonians at a much earlier time had quite sophisticated knowledge of at least some aspects of number theory. At present we do not know how thoroughly they could prove some of their results, but perhaps further archaeological work will yield that information.

Whatever the extent to which the Babylonians carried their proofs, we are certain that the Greeks were very thorough in theirs. Strangely enough, the original works of the Greeks up to about 300 B.C. have almost completely disappeared. One of the main reasons for this disappearance is a remarkable one: In Alexandria (a Greek city in Egypt) a mathematician named Euclid wrote a truly monumental work compiling all of the important mathematical results that had been proved up to his time. Euclid's *Elements* was so thorough and so well written that it superseded all the previously written mathematical works of the Greeks. As a result, no care was taken to preserve and copy the earlier works, so today very little remains of them. Euclid's *Elements* contains not merely number theory and geometry, but *all* aspects of mathematics known at that time. Some of the notions are actually related to calculus.

Why was the theory of natural numbers of such great interest to early mathematicians? One reason is to be found in its connections with geometry (discussed in Chapter 3), while another reason, very powerful at that time, was the belief in magic powers possessed by natural numbers. For instance, seven has been a favorite "lucky" number of many civilizations. There are seven days to a week, seven days to the Biblical creation, seven years bad luck for breaking a mirror, seven choirs of angels, and altogether perhaps a hundred favored uses of the number seven in the New Testament alone.

To take another example, the number six has the property of being equal to the sum of its proper divisors, one, two, and three. A number with this property is said to be "perfect," and such numbers were once believed to have marvelous powers. St. Augustine believed that God's work of creation was done over a span of six days because both the number six and God's work are perfect. (After 6, the next larger perfect number is 28, and the one following 28 is 496. So far, there are just 23 perfect numbers known, and all of them are even.)

Today, of course, this element of magic in numbers is no longer so appealing. But if we have lost the taste for magic, we have probably gained in plain curiosity, so we still have good reason to spend time investigating the properties of the natural numbers which so intrigued the early mathematicians.

EXERCISES

1. Find at least three numbers other than seven that seem to you to have been "favorite" numbers in times past and present. For each of your three numbers, give one or more contexts in which it is used as a favorite number.

2. For each of the numbers $n = 2, 3, 4, 5, \cdots, 48, 49, 50$, find the sum of the proper divisors of n. (A natural number d is called a proper divisor of n if d is less than n and n/d is a natural number.) List those values for n for which the sum of the proper divisors is less than n; these used to be called *deficient* numbers. List those values of n for which the sum of the proper divisors is greater than n; these used to be called *abundant* numbers.

3. Look at the divisor sums of the numbers 2, 4, 8, 16, and 32 in Exercise 2. What would you guess to be the sum of all proper divisors of $2^n (= 2 \times 2 \times \cdots \times 2$, n times)?

4. From the year 1860 to the present, six presidents of the United States have died while in office. Tell who the six were, and tell the year in which each one was first elected president. (This striking coincidence in the dates of election is mentioned very frequently.)

5. Look up or take a guess at the reason the number 13 came to be considered unlucky.

6. "Here is wisdom. Let him that hath understanding count the number of the beast: for it is the number of a man; and his number is Six hundred threescore and six." (Revelation 13, 18) The reference, as well as scholars can determine, seems to have been to Nero. Each letter of the Hebrew alphabet has a numerical significance, and the sum of the values of the letters QSR NRWN (meaning "Caesar Nero") is 666. There have been attempts to associate the number 666 with other figures in history, always in order to show that Mr. 666 is the beast "having seven heads and ten horns." Find out the names of at least two historical figures who have been honored by their contemporaries with the number 666. In each case, tell how the association was made.

7. Explain how gematriya is done. (Suggested reference: Chapter 7 of *The Chosen* by Chaim Potok. Simon and Schuster, New York, 1967.)

8. Show that any number whose last two digits are 00 is abundant. [*Hint:* Write the number as something times 100 and then use the various proper divisors of 100 to construct several large proper divisors of the original number.]

9. Write down an unending list of abundant numbers, all different. Write down an unending list of deficient numbers, all different. [*Hint:* Use Exercises 8 and 3.]

2. Primes, Composites, and the Infinitude of the Composites

In studying the natural numbers, one of the first steps the early mathematicians took was to split all the natural numbers into two classes according to a divisibility criterion. Following their lead, we make the two definitions:

DEFINITION. If a natural number n is divisible by a natural number other than itself and 1, we will call n a *composite number*.

DEFINITION. If a natural number n is divisible only by itself and 1, we will call n a *prime number*.[1]

For example, among the first thirty natural numbers, the composite numbers are 4, 6, 8, 9, 10, 12, 14, 15, 16, 18, 20, 21, 22, 24, 25, 26, 27, 28, and 30, while the prime numbers are 1, 2, 3, 5, 7, 11, 13, 17, 19, 23, and 29. Every one of the first thirty natural numbers is to be found in one or the other of the above two lists. The reason is that we defined composites and primes to be opposites of each other; every natural number is either a composite number or a prime number.

According to the definitions, the composites are those natural numbers that can be "broken down" into products of smaller natural numbers, while the primes are those natural numbers that cannot be so broken down. We might say that the primes are the "atoms" of number theory.

If the primes are the atoms, then the composites are the "molecules" of number theory, because every composite number can be written as the product of prime numbers. For example, $100 = 2 \cdot 2 \cdot 5 \cdot 5$, $39 = 3 \cdot 13$, and $42 = 2 \cdot 3 \cdot 7$.

In order to write down a composite number n as a product of primes, one goes through a factorization process. Because n is composite, we have

$$n = a \cdot b,$$

where a and b are both natural numbers smaller than n. If either or both of the factors a and b are composite, write each composite factor as a product of two smaller factors. If any of the new factors are composite, factor them, and continue factoring as long as a composite factor remains. This factorization process must eventually come to an end, because each time a composite factor is factored further, it is replaced with two numbers smaller than itself. When the factorization process does end, of course, n is written as a product of unfactorable numbers, or primes.

For example, to write 100 as a product of prime numbers, we could factor as follows:

---

[1] According to this definition, the number 1 is a prime number. Some other books consider the number 1 to be neither prime nor composite.

$$100 = 10 \cdot 10$$
$$= 2 \cdot 5 \cdot 2 \cdot 5,$$

or as follows:

$$100 = 5 \cdot 20$$
$$= 5 \cdot 5 \cdot 4$$
$$= 5 \cdot 5 \cdot 2 \cdot 2,$$

or as follows:

$$100 = 4 \cdot 25$$
$$= 2 \cdot 2 \cdot 5 \cdot 5,$$

or in several other ways.

We mentioned that the notions of prime and composite are exact opposites of each other. That is, each natural number is either a prime number or a composite number (and cannot be both prime and composite). Therefore, the collection of all the natural numbers may be split into two smaller collections — the collection of primes and the collection of composites.

Now whenever an infinite collection (like the collection of all natural numbers 1, 2, 3, 4, $\cdots$) is split into two smaller collections, at least one of the smaller collections must also be infinite. (For if both smaller collections were finite, then the larger collection would have finite size equal to the sum of the sizes of the two smaller collections.)

In the case of the natural numbers, which of the two smaller collections, the composites and the primes, are infinite? It turns out that they are both infinite, and we will prove that is the case. To prove that the composite numbers form an infinite collection is very easy:

THEOREM 1. There are infinitely many composite natural numbers.

Proof. We will exhibit an unending list of natural numbers, every one of which is composite. Since the collection of all composite numbers will contain all the numbers in this list, it will certainly be infinite. There are many ways to exhibit such a list, but perhaps the simplest is to write down in order all the even numbers larger than 2: 4, 6, 8, 10, 12, $\cdots$. Any even number larger than 2 is divisible by at least one number other than itself and 1, namely, by the number 2, and so does satisfy the definition of a composite number.

For a simple exercise, you might like to devise other unending lists of composite numbers. It is remarkable how many completely different ways there are in which to do this. Let us look at one other way.

DEFINITION. If n is any natural number, let n *factorial*, written $n!$, be the number

$$n(n-1)(n-2) \cdots 3 \cdot 2 \cdot 1.$$

For example, for the values $n = 1, 2, 3, 4,$ and 5, we have

$$1! = \qquad\qquad 1 = \quad 1$$
$$2! = \qquad\quad 2 \cdot 1 = \quad 2$$
$$3! = \qquad 3 \cdot 2 \cdot 1 = \quad 6$$
$$4! = \quad 4 \cdot 3 \cdot 2 \cdot 1 = \quad 24$$
$$5! = 5 \cdot 4 \cdot 3 \cdot 2 \cdot 1 = 120.$$

By the definition, $n!$ is always divisible by n. Therefore, for any value of n larger than 2, $n!$ is a composite number, since n is then different from both $n!$ and 1. Thus the list $3!, 4!, 5!, 6!, \cdots$ is an unending list of composite numbers.

The factorial is a rather sophisticated gadget to be used merely for obtaining an unending list of composite numbers. Our real reason for introducing it here is to have it available for later use. It arises in a variety of places, inside and outside of number theory. Our first major use of it will be in the next section.

EXERCISES

1. Factor each of the composite numbers less than 31 into a product of prime numbers.

2. Factor each of the following numbers into a product of prime numbers:

    512   198   196   105   10,000.

3. Construct at least three more unending lists of composite numbers (thus giving three more proofs of Theorem 1).

4. A tongue-in-cheek mathematics article published a few years ago listed a book entitled *A Short Table of Even Primes*. Assuming you have a friend who does not understand the joke, write an explanation for his benefit. How long is the table of even primes?

2. PRIMES, COMPOSITES 21

5. The instructions of this exercise give a method, called the *sieve of Eratosthenes*, for separating primes from composites. Write down the natural numbers in order: 1, 2, 3, 4, $\cdots$ up to some number n. Circle 1 to mark it as a prime. Circle 2 to mark it as a prime, and then cross out every second number from then on as composite (because divisible by 2). Find the next un-crossed, uncircled number to the right, 3, circle it to mark it as a prime, and then cross out every third number from then on as composite (because divisible by 3). Find the next uncrossed, uncircled number to the right, 5, circle it to mark it as a prime, and then cross out every fifth number from then on as composite (because divisible by 5). And so on. At the end, all the primes in the list will be circled and all the composites will be crossed out. Use Eratosthenes' sieve to find all primes less than 250. [*Historical note:* Two centuries before the birth of Christ, Eratosthenes calculated the circumference of the earth. His answer was within about 5 percent of the value we know today.]

6. In the series of natural numbers up to and including 250, what is the longest string of consecutive composite numbers?

7. In the series of natural numbers up to and including 250, what is the longest string of consecutive prime numbers? Could there ever be a longer string farther out in the natural numbers? Why?

8. What fraction of the natural numbers 1, 2, 3, $\cdots$, 50 are primes? Answer the same question with "50" replaced by: 100, 150, 200, 250. Would you be willing to make any guess on the basis of these five values?

9. Prove that the sieve of Eratosthenes really works. [*Hint:* The first task is to realize that there is something to be proved. It is quite clear that every number crossed out is composite. What is not obvious is that every composite winds up being crossed out or, equivalently, that every circled number is prime.]

10. Compute the values of

    6!   7!   8!   9!   10!

11. Without doing any computations, tell why the numbers $11! + 1$, $12! + 1$, $13! + 1$, $\cdots$, $20! + 1$ are all odd numbers.

12. Without doing any computations, tell why none of the numbers listed in Exercise 11 are divisible by 3. [*Hint:* You can actually tell what remainder each leaves when divided by 3.]

3. The Infinitude of the Primes

You will recall we mentioned in the last section that both classes of natural numbers, composites and primes, are infinite. Proving that the composites are infinite is easy, since there are many ways in which to write down unending lists of composite numbers.

Because composites are so easy to find, we might expect that their opposites, the primes, will be hard to find, tucked in as they are amongst all the composites. To prove that there are infinitely many primes, then, we should not expect to be able to use such a simple device as exhibiting an unending list of numbers, all primes. The reasoning used in proving Theorem 1 could have occurred to almost anyone, once he had decided to prove such a result. In proving the infinitude of the primes, however, the reasoning is not so straightforward. Almost anyone can *follow* it as it is performed, but as he follows it he might well doubt if he could have discovered or invented the series of steps taken. Whoever the inventor was (some Greek mathematician prior to Euclid), he deserves credit for one of the first great discoveries in number theory.

THEOREM 2. There are infinitely many prime natural numbers.

Proof. We will show that for each prime p there is at least one prime larger than p. That is, there is no such thing as a *largest prime*, so the collection of all prime numbers must be infinite.

Consider the number $p! + 1$. When this number is divided by any number larger than 1 and less than or equal to p, it leaves a remainder of 1. Thus $p! + 1$ is *not* divisible by any number less than or equal to p except the number 1. Therefore any prime larger than 1 which divides $p! + 1$ must be larger than p. Consequently, all we need to complete the proof is to show that $p! + 1$ always has a prime divisor larger than 1.

In case $p! + 1$ is prime, it itself is such a prime divisor. In case $p! + 1$ is composite, factor it into a product of prime numbers larger than 1. Pick any one of these prime factors d of $p! + 1$. Because d divides $p! + 1$, it is larger than p. The proof is complete.

It is interesting to study what this proof does *not* do, as well as what it does. For each prime p, it does produce some prime larger than p. It does not necessarily produce the very next prime after p. (For example, $3! + 1 = 7$, a prime but not the next prime after 3.) In fact, to this day there is no practical way known for finding from the value of p the value of the next larger prime. It was the inventor's

3. THE INFINITUDE OF THE PRIMES

genius that he did not try to prove more than he needed. All that was required was to prove the existence of *some* prime larger than p, and this he did prove.

Depending on the value of p, $p! + 1$ may be prime, or it may be composite. (For example, $3! + 1 = 7$ is prime, but $5! + 1 = 121 = 11 \cdot 11$ is composite.) Therefore the method of the proof does not even give an explicit formula for *some* prime larger than p. In certain cases the prime will be given by the formula $p! + 1$, while in other cases it will be necessary to factor the number $p! + 1$ in order to obtain a prime larger than p.

EXERCISES

1. For which values of p: 1, 2, 3, 5, and 7 is $p! + 1$ composite and for which values is it prime? In which of these five cases is one of the prime divisors of $p! + 1$ the next prime larger than p?

2. How many pairs of primes p and q are there such that $q - p = 3$? Prove your answer.

3. Exercise 2 has a simple answer, but if we change the number 3 to the number 2, we obtain a famous unsolved problem: How many pairs of primes p and q are there such that $q - p = 2$? Such pairs are called "twin primes," and it is suspected very strongly that there are infinitely many pairs of twin primes. List all pairs of twin primes, both members of which are less than 250. [*Suggestion:* Use the list of primes produced by Eratosthenes' sieve in Exercise 5 of the preceding section.]

4. How many triplets of primes p, $p + 2$, $p + 4$ are there? Prove your answer. [*Hint:* Show that one of p, $p + 2$, $p + 4$ is divisible by 3.]

5. If p_k is the kth number in the list of primes 1, 2, 3, 5, 7, $\cdots$, rewrite the proof of Theorem 2 using the number $(1 \cdot 2 \cdot 3 \cdot 5 \cdot 7 \cdot \ldots \cdot p_k) + 1$ instead of the number $p_k! + 1$. (The original Greek proof of the infinitude of the primes used this number rather than $p_k! + 1$.)

6. Write every even number less than 100 as a sum of two primes. (It was conjectured by the German mathematician Christian Goldbach (1690–1764) that every even number is the sum of two primes. No proof of Goldbach's conjecture has ever been found; it has been verified individually for all even numbers less than 33,000,000.)

7. Corresponding to the guess that every even number is the sum of two primes, make a guess about representing odd numbers as sums of primes. Show that if Goldbach's conjecture is true, then your guess must also be true. [*Hint:* Subtracting an odd prime from an odd number produces an even number.]

8. If you are given a single natural number n to be tested to see if it is prime, about the most economical way to perform the test is to try dividing it by the natural numbers 2, 3, 4, 5, $\cdots$. However, it is not necessary to try as a divisor every number up to n. At what point, if no exact division has yet occurred, can one conclude that n is prime?

9. Suppose n is a natural number, say greater than 10. Is $n! + 2$ prime or composite? What about $n! + 3$, $n! + 4$, and so on? For what value of k can you no longer be certain of the primality or compositeness of $n! + k$?

10. Use your answers to Exercise 9 to prove that for every natural number n, there is some string of n consecutive natural numbers, all composite. (Yes, *somewhere*, far out in the list of natural numbers, there are 1,000,000,000,000,000 consecutive natural numbers, not a single one of them prime.)

11. Invent at least two numbers other than the one in Exercise 5 that can be used in the proof of Theorem 2 in place of $p! + 1$.

4. Prime-Generating Formulas

In this section we will discuss different attempts to invent formulas that produce only prime numbers. The motivation behind such attempts stems in part from the observation that it is very easy to make up formulas that always give composite numbers. For example, the list of all even numbers greater than two that we used in the proof of Theorem 1 can be obtained from the formula $2n + 2$ by setting $n = 1, 2, 3, 4, 5$, and so on.

Can we find a formula so simple, or nearly so, that gives nothing but primes when we substitute the numbers 1, 2, 3, 4, 5, $\cdots$ for the unknown? The polynomial formula

$$n^2 - n + 41$$

produces primes for the first forty values of n:

$$n = 1, 2, 3, 4, 5, \cdots, 38, 39, 40.$$

4. PRIME-GENERATING FORMULAS

However, you can see very easily that when $n = 41$ it finally yields a composite number, with 41 as a divisor:

$$41^2 - 41 + 41 = 41^2 = 41 \cdot 41.$$

Expressions like:

$$n^2 - n + 1 \qquad 3n^5 - 4n^4 + n - 5$$
$$10n \qquad 35$$
$$n^3 + 23 \qquad n^5$$

all go by the name of polynomials. A polynomial is defined to be any sum of terms having the form cn^k, where c can be any number and k can be either 0 or any natural number. The symbol n (or x or whatever symbol is used) is called the *unknown* or the *variable*. The highest value of the exponent k appearing in a polynomial is called the *degree* of the polynomial. For instance, the polynomial $n^3 + 23$ has degree 3, the polynomial $10n$ has degree 1, and the polynomial 35 has degree 0.

Polynomials were among the first formulas to be studied in mathematics, so it was natural to inquire if there could be any polynomial formula that would produce only primes when the values 1, 2, 3, 4, 5, $\cdots$ were substituted for the unknown n. The answer is no, unless the polynomial is simply a prime number to start with. In fact, every polynomial of degree 1 or greater with integer coefficients must *infinitely often* take on composite values as the unknown assumes the values 1, 2, 3, 4, 5, $\cdots$ in succession. For example, $n^2 - n + 41$ is composite for every natural number n which is a multiple of 41: 41, 82, 123, 164, and so on.

The Swiss mathematician Leonhard Euler (1707–1783) discovered the peculiarities of the polynomial $n^2 - n + 41$ that we have just discussed. Of course, he realized that it gave composite numbers for some values of n, but he was intrigued by the fraction of time (slightly under $\frac{1}{2}$ as far as it has been checked out) that it gave primes. It is still somewhat mysterious today. It gives composites infinitely often, but no one knows if its performance of giving primes is sustained as n gets very large. Perhaps from some point on it turns around and gives nothing but composites. No one knows.

Raising a number to a variable power is another means of writing down a variety of simple-looking formulas. For example, let us investigate the expression

$$2^n + 1.$$

Table 1 shows how this formula behaves when the first eight natural numbers are substituted for n.

TABLE 1. Factorization of $2^n + 1$ for the first eight values of n.

n	$2^n + 1$	Factorization?
1	3	prime
2	5	prime
3	9	$3 \cdot 3$
4	17	prime
5	33	$3 \cdot 11$
6	65	$5 \cdot 13$
7	129	$3 \cdot 43$
8	257	prime

We see that for these first eight values of n, the formula produces primes when n is a power of 2: 1, 2, 4, and 8, but it fails for the other values of n: 3, 5, 6, and 7. We are led, then, to ask two questions. First, is it true that $2^n + 1$ is always composite when n is not a power of 2? Second, is it true that

$$2^{2^k} + 1$$

is always prime? The answer to the first question is given by Theorem 3 below.

THEOREM 3. If n is any natural number other than a power of 2, then $2^n + 1$ is composite.

Proof. If n is not a power of 2, then we can factor n into a product $n = a \cdot b$, where the factor b is an odd number greater than 1. Now, whenever b is odd, the polynomial $x^b + 1$ can be factored:

$$x^b + 1 = (x + 1)(x^{b-1} - x^{b-2} + x^{b-3} - \cdots - x + 1).$$

(For those not familiar with this factorization, it can be verified on-the-spot by multiplying the right-hand side of the equation, as below

$$\begin{array}{r} x^{b-1} - x^{b-2} + x^{b-3} - \cdots - x + 1 \\ x + 1 \\ \hline x^{b-1} - x^{b-2} + x^{b-3} - \cdots - x + 1 \\ x^b - x^{b-1} + x^{b-2} - x^{b-3} + \cdots + x \\ \hline x^b \qquad\qquad\qquad\qquad\qquad\qquad + 1 \end{array}$$

obtaining the polynomial on the left-hand side.)

We can use this factorization of $x^b + 1$ to obtain a factorization of the number $2^n + 1$. In place of the variable x, substitute the value 2^a:

$$\begin{aligned} 2^n + 1 &= 2^{a \cdot b} + 1 \\ &= (2^a)^b + 1 \\ &= (2^a + 1)((2^a)^{b-1} - (2^a)^{b-2} + \cdots - 2^a + 1). \end{aligned}$$

Since b is greater than 1, we know that a is less than n, so the number $2^a + 1$ is a divisor of $2^n + 1$ which is smaller than $2^n + 1$ and, of course, larger than 1. Therefore $2^n + 1$ is composite. This completes the proof.

This theorem was known to Pierre de Fermat (1601–1665), who is considered to be the first great number theorist since the time of the Greeks. By profession, he was a jurist; he did his mathematics as a hobby, but he did some truly wonderful things in it. Knowing that numbers of the form $2^n + 1$, n not a power of 2, are composite, he was led to the suspicion that the others, $2^{2^k} + 1$, might all be primes. We call these numbers, $2^{2^k} + 1$, Fermat numbers in honor of his study of them, but it turned out that his guess was wrong. The next Fermat number beyond those in our table, $2^{2^4} + 1$, is a prime also, but the one immediately following that, $2^{2^5} + 1$, was shown by Euler to be composite. Its factorization into a product of primes is

$$2^{2^5} + 1 = 641 \cdot 6{,}700{,}417.$$

Since Euler's time, various mathematicians have established that $2^{2^k} + 1$ is composite also for the values $k = 6, 7, 8, \cdots, 15, 16$, and it is known to be composite for certain larger values of k as well. There is not presently known any value of k other than 0, 1, 2, 3, and 4 for which the Fermat number $2^{2^k} + 1$ is prime. However, one would hate to hazard any guesses on the basis of existing evidence. And since the Fermat numbers increase in size so rapidly as k increases ($2^{2^{16}} + 1$ is 19,729 digits long), it is unlikely that very much more evidence will be available soon. The Fermat numbers may well remain mysterious for a long time to come.

We have now spent some time examining two remarkable formulas:

$$n^2 - n + 41$$
$$2^{2^k} + 1.$$

Despite their high promise at the beginning, the verdict in each case was that the expression does not produce only prime numbers. To these two "unsuccessful" prime-generating formulas, we can add the formula

$$p! + 1.$$

This last expression does just what the Greeks wished; it shows that there is a prime larger than p, but the expression itself is not necessarily prime. To get at the prime larger than p, we may have to factor the number $p! + 1$.

To this day, no one knows of a useful formula in variable n which will give only primes for the values $n = 1, 2, 3, 4$, and so on. We remarked earlier that no polynomial with integer coefficients and degree at least 1 can yield only primes, but there are an infinity of other possibilities which have never been fully explored. However, most mathematicians are doubtful that any useful prime-generating formula will ever be found.

That vague adjective "useful" is necessary, because there are some artificial-looking formulas that do give all the primes, in increasing order, when the numbers $1, 2, 3, 4, 5, \cdots$ are inserted for the variable. The artificiality lies in the fact that each such formula contains some constant that cannot be completely known unless one already knows all the primes. It is as though the primes are recorded on this single constant as music is recorded on magnetic tape. The substitution of the numbers $1, 2, 3, \cdots$ into the formula causes the primes to be "played" back, one after the other.

One such formula, which we are writing down only so you can see what it looks like, is

$$[10^{2^n} c] - 10^{2^{n-1}} [10^{2^{n-1}} c].$$

The constant c is the number

$$c = .0102000300000005000000000000000700 \cdots.$$

The rule for forming c is that the nth prime number is inserted into the decimal expansion of c so that the rightmost digit is in the 2^n place following the decimal point. The square brackets denote the "integral part" of whatever appears inside them. For instance, $[21.347]$ means the number 21, and $[9.99]$ means the number 9.

EXERCISES

1. Show that for each natural number which is a multiple of 41, $n^2 - n + 41$ is a multiple of 41.

2. Find a value of n for which $n^2 - n + 17$ is composite. Factor $n^2 - n + 17$ into a product of primes for this value of n.

3. Verify that $n^2 - n + 17$ is prime for $n = 1, 2, 3, \cdots, 14, 15, 16$.

4. Find an unending list of values for n which $n^2 - n + 17$ is composite. Prove your answer is right.

5. Let c be a natural number greater than 1. Prove that $n^2 - n + c$ is composite for infinitely many values of n.

6. Factor each of the numbers below into a product of primes:
$$2^9 + 1 \qquad 2^{10} + 1 \qquad 2^{12} + 1.$$
[*Hint:* Use the factorization of $x^b + 1$ for b odd.]

7. Evaluate the number $2^{16} + 1$. Estimate how long it would take you with pencil and paper to verify that this number is prime. Explain how you arrived at your estimate.

8. Is it possible to factor $x^b + 1$ as
$$x^b + 1 = (x + 1)(x^{b-1} - x^{b-2} + x^{b-3} - \cdots - x + 1)$$
when b is even? Demonstrate with $x^8 + 1$.

9. Is it possible to factor $x^b + 1$ *at all* when b is even? [*Hint:* Try $x^6 + 1$ before jumping to conclusions.]

10. Verify that
$$2^{2^5} + 1 = 641 \cdot 6{,}700{,}417.$$

11. Show that the formula $4n + 3$ yields composite numbers for infinitely many natural numbers n. [*Hint:* Write down an unending list that does the job.]

12. Show that the formula $4n + 3$ yields prime numbers for infinitely many natural numbers n. [*Hint:* If p is a prime of the form $4n + 3$, show that some prime factor of $4(p!) - 1$ is again of the form $4n + 3$; the bulk of the proof is the same as that of Theorem 2.]

13. Use the results of your work with the sieve of Eratosthenes or use a table of primes to answer the following questions. Of the natural numbers less than 250, how many are primes of the form $4n + 3$? of the form $4n + 2$? of the form $4n + 1$? of the form $4n$? Would you be willing to hazard a guess as to whether or not there are an infinite number of primes of the form $4n + 1$?

5. The World's Largest Prime

There is, of course, no such thing as *the* largest prime, as we showed in the proof of Theorem 2. However, as of this writing, there is a largest natural number which is known for certain to be prime. That number, shown in the figure, is 3376 digits long when written in decimal form. As you might guess from the postmark, the discovery was made by a mathematician (Donald Gillies) at the University of

Illinois. The date of the discovery was 1963, somewhat earlier than the date on the stamp. How did anyone become interested in such a monster number, and how did anyone guess that certain numbers of its general shape could be shown to be prime? In this section we will answer this double-barreled question.

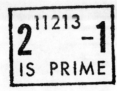 

You will recall that in the previous section we investigated what condition on n was needed for a number of the form

$$2^n + 1$$

to be prime. The answer was that it was *necessary* for n to be a power of 2, but it turned out that not all numbers of the form

$$2^{2^k} + 1$$

were prime.

It does not require all that much curiosity (if you are a mathematician) to ask the same question about numbers of the form

$$2^n - 1.$$

What condition on n is needed for $2^n - 1$ to be prime? The answer is that n itself must be a prime. The answer this time is quite a bit different from the answer obtained the previous time, but the argument leading to the answer is very much the same. The argument this time involves factoring the polynomial $x^b - 1$:

$$x^b - 1 = (x - 1)(x^{b-1} + x^{b-2} + x^{b-3} + \cdots + x + 1).$$

A series of exercises at the end of this section will lead you through the proof that n must be prime in order for $2^n - 1$ to be prime.

Not all cases when n is a prime will yield $2^n - 1$ prime, however. For example,

$$2^{11} - 1 = 2049 = 23 \cdot 89.$$

Numbers of the form $2^p - 1$, p a prime number larger than 1, are called *Mersenne numbers*, after Marin Mersenne (1588–1648), who made them a specialty of his. (Mersenne, like his colleague Fermat, was an amateur mathematician, although not nearly as brilliant. By profession, he was a Franciscan priest.) The original interest in Mer-

5. THE WORLD'S LARGEST PRIME

senne numbers lay in their connection with perfect numbers. It turns out that an even number is perfect if and only if it is of the form

$$2^{p-1}(2^p - 1),$$

where $2^p - 1$ is a Mersenne number which is prime.

Now Mersenne numbers are of other interest, ever since a mathematician named Edouard Lucas (1842–1891) found a method for testing primality of Mersenne numbers with far fewer arithmetical operations than are presently needed for other numbers. Thus, Lucas' method has given us a way of finding a few very large primes, by testing large Mersenne numbers to see if they are prime. The largest number which has been found to be prime by Lucas' method is $2^{11213} - 1$.

To test this number for primality by test-dividing it by every natural number less than its square root (assuming optimistically that each such fantastic division could be performed in one-millionth of a second) would require more years than the earth is old. Lucas' test reduces the number of needed calculations to an amount that a high-speed computer can handle in a matter of hours.

All numbers $2^p - 1$ for p a prime less than 12,000 have now been tested for primality. The complete[1] list of the exponents p less than 12,000 for which the Mersenne number $2^p - 1$ is prime is:

2, 3, 5, 7, 13, 17, 19, 31, 61, 89, 107,
127, 521, 607, 1279, 2203, 2281, 3217,
4253, 4423, 9689, 9941, 11213.

Corresponding to these 23 primes, there are now 23 even perfect numbers known. It is an unsolved problem as to whether there are infinitely many prime Mersenne numbers (and thus infinitely many even perfect numbers). As we mentioned earlier, no odd perfect number has ever been found.

EXERCISES

1. Write the perfect numbers 6 and 28 in the form $2^{p-1}(2^p - 1)$.
2. Write the following numbers in decimal form:

    $2^4(2^5 - 1) \qquad 2^6(2^7 - 1) \qquad 2^{12}(2^{13} - 1).$

3. Find all proper divisors of the number $2^4(2^5 - 1)$ and then add them, thus verifying that $2^4(2^5 - 1)$ is a perfect number.
1. A 24th Mersenne prime was discovered almost simultaneously with the publication of this book. Its exponent p is 19937.

4. Prove that when b is greater than 1,

$$x^b - 1 = (x - 1)(x^{b-1} + x^{b-2} + x^{b-3} + \cdots + x + 1)$$

by multiplying the two polynomials on the right-hand side.

5. Show that if $n = a \cdot b$ is composite, then $2^n - 1$ is divisible by $2^a - 1$.

6. Conclude, therefore, that if $2^n - 1$ is prime, then n must be prime.

7. Suppose g is an odd number and n is greater than 1. Is $g^n - 1$ ever a prime? [*Hint:* This was meant to be easy.]

8. Name two different divisors of $2^{15} - 1 = 32{,}767$, both greater than 1 and less than 32,767.

9. Factor 32,767 into a product of primes.

10. Prove that if p and q are odd primes larger than 1, then the odd number $p \cdot q$ cannot be perfect. [*Hint:* Show, in fact, that the sum of the proper divisors of $p \cdot q$ is less than $p \cdot q$. That is, $p \cdot q$ is a deficient number.] What if one or both of p and q are equal to 1?

11. If you divide a Mersenne number $2^p - 1$ by 4, what is the remainder?

12. Suppose the result stated in Exercise 12 of the previous section had been false. What would that, together with Exercise 11 of this section, have said about the number of Mersenne numbers which are prime?

13. If $2^p - 1$ is a prime Mersenne number, describe the collection of all proper divisors of $2^{p-1}(2^p - 1)$. Prove that their sum is equal to $2^{p-1}(2^p - 1)$.

14. If $2^p - 1$ is not prime, is $2^{p-1}(2^p - 1)$ deficient or abundant? (See Exercise 2, Section 1 for the definitions of deficient and abundant.)

15. Use the method of proof of Theorem 2 to write down a number all of whose prime divisors except 1 are larger than $2^{11213} - 1$.

---

6. Euclid's Algorithm

By now you must have noticed that finding the divisors of a large natural number can be very time-consuming. It may come as a surprise

6. EUCLID'S ALGORITHM

then to learn that the process of finding out whether two natural numbers have any divisors in common is relatively easy and not at all lengthy. In fact, it is fairly easy to calculate a particular kind of common divisor of any two natural numbers:

DEFINITION. If a and b are two natural numbers, the *greatest common divisor* of a and b, written (a, b), is the largest natural number that divides both a and b.

The method for finding the greatest common divisor appears in Euclid's *Elements*, and as a result it has been given the name *Euclid's algorithm*, although it may well have been the discovery of an earlier mathematician. There is one key idea behind Euclid's algorithm, and that key idea involves what are called linear combinations:

DEFINITION. Let a and b be natural numbers. Any integer which is equal to $ia + jb$, where i and j are integers (not necessarily positive), is said to be a *linear combination* of a and b.

For example, let $a = 3$ and $b = 7$. Then some numbers which are linear combinations of 3 and 7 are

$$10 = 1 \cdot 3 + 1 \cdot 7 \qquad 25 = 6 \cdot 3 + 1 \cdot 7$$
$$1 = -2 \cdot 3 + 1 \cdot 7 \qquad 0 = 0 \cdot 3 + 0 \cdot 7$$
$$-2 = 4 \cdot 3 - 2 \cdot 7.$$

Key Idea

If we divide one natural number y by another natural number x, and if both x and y are linear combinations of a and b, then the remainder is also a linear combination of a and b.

Proof. Let x be the linear combination $ia + jb$, and let y be the linear combination $ka + lb$. Let q be the quotient and r the remainder when y is divided by x:

$$x \overline{)y} \begin{array}{c} q \quad r \\ \end{array} .$$

By the definition of the division process,

$$y = qx + r$$

or

$$ka + lb = q(ia + jb) + r.$$

Solving this second equation for r, we have

$$r = ka + lb - q(ia + jb)$$
$$= (k - qi)a + (l - qj)b$$

and this completes the proof.

The above proof went so quickly that it is well worth seeing a replay of the argument applied step-by-step to an example. We saw before that $10 = 1 \cdot 3 + 1 \cdot 7$ and $25 = 6 \cdot 3 + 1 \cdot 7$ are both linear combinations of 3 and 7. If we divide 25 by 10:

$$10 \overline{)25} \begin{array}{c} 2 5 \\ \end{array},$$

we have a quotient of 2 and a remainder of 5. By definition of the division process,

$$25 = 2 \cdot 10 + 5.$$

In terms of the linear combinations, this last equation reads

$$6 \cdot 3 + 1 \cdot 7 = 2 \cdot (1 \cdot 3 + 1 \cdot 7) + 5.$$

Now, just as we did in the proof itself, we can solve this equation for the remainder 5:

$$5 = 6 \cdot 3 + 1 \cdot 7 - 2 \cdot (1 \cdot 3 + 1 \cdot 7)$$
$$= (6 - 2) \cdot 3 + (1 - 2) \cdot 7$$
$$= 4 \cdot 3 - 1 \cdot 7.$$

You see, we have pushed the numbers 10 and 25 through the proof of the key idea, and the result was the remainder 5 written as a linear combination of 3 and 7. The proof furnishes a genuinely practical method for writing the remainder as a linear combination of the numbers a and b.

We will now use this key idea to prove a theorem about the greatest common divisor, and the theorem in turn will be our guide for developing Euclid's algorithm.

THEOREM 4. The greatest common divisor (a, b) of two natural numbers a and b is equal to the smallest natural number which is a linear combination of a and b. If c is a divisor of both a and b, then c divides (a, b).

Proof. Let $x = ia + jb$ be the smallest natural number which is a linear combination of a and b. Let $y = ka + lb$ be any natural number which is a linear combination of a and b. Divide y by x. According to the key idea, the remainder is again a linear combination of a and b. This forces the remainder to be 0, since if it were nonzero, it would be a natural number linear combination of a and b which is *smaller* than the divisor x, and that is contrary to our choice of x. *Therefore*, x divides any natural number y which is a linear combination of a and b.

In particular, x divides the linear combinations

$$1 \cdot a + 0 \cdot b = a$$

and

$$0 \cdot a + 1 \cdot b = b.$$

That is, x is a *common divisor* of a and b.

Now let c be any divisor of both a and b. Then c also divides the linear combination

$$i \cdot a + j \cdot b = x.$$

Therefore x is greater than or equal to every common divisor of a and b. That is, x is the greatest common divisor (a, b) of a and b. In the course of proving this, we have proved that any common divisor c of a and b will always divide $x = (a, b)$, which was the second assertion of the theorem. The proof is complete.

We have now identified the greatest common divisor as a very special kind of linear combination. What Euclid's algorithm does, in turn, is furnish a sure-fire procedure by which this particular linear combination can be calculated.

Let us refer back to the key idea, thinking of it now as giving a method for creating a new linear combination of a and b from two old ones. The new linear combination is obtained as a remainder after division, and by definition a remainder is a smaller number than the divisor. If the smaller of the two linear combinations is used as the divisor, we see that the key idea gives a method for creating a new, smaller linear combination from two old ones. For example, we saw already that division of the number 25 by the number 10 produced the remainder 5, and all three of these numbers were linear combinations of the numbers 3 and 7. The new linear combination, being smaller, is closer to being (a, b) than either of the original two linear combinations (unless the new combination is zero!). This is the philosophy that leads us to consider the series of divisions which make up Euclid's algorithm.

Suppose we wish to find the greatest common divisor $(32, 50)$ of the two natural numbers 32 and 50. We wish to apply the key idea over and over in order to get smaller and smaller linear combinations of 32 and 50. Hopefully, we will eventually obtain a linear combination which is so small that it is the smallest natural number linear combination of 32 and 50, namely, the greatest common divisor of these two numbers.

To begin, we must pick two linear combinations of 32 and 50 and divide the one by the other. The very simplest linear combinations are the ones right before our eyes, 32 and 50 themselves. Why not use them? If we divide the larger by the smaller

$$\begin{array}{r} 118 \\ 32\overline{)50} \end{array},$$

we obtain a remainder, 18, which is smaller than both 32 and 50. By the key idea, this remainder is a linear combination of 32 and 50.

Now, if we take the two linear combinations 18 and 32 and divide the larger by the smaller

$$\begin{array}{r} 114 \\ 18\overline{)32} \end{array},$$

we obtain a remainder which is smaller than either of them. This remainder, 14, is again a linear combination of 32 and 50, on account of the key idea. Let us now continue this process without using any more words.

$$\begin{array}{r} 14 \\ 14\overline{)18} \\ 32 \\ 4\overline{)14} \\ 20 \\ 2\overline{)4} \end{array}$$

Each one of the remainders, 18, 14, 4, 2, and 0, is a linear combination of the original two numbers 32 and 50. The key idea assures us that this is the case. Because of the way in which we performed the divisions, the remainders kept decreasing as we went along. That is, they became closer and closer to the smallest natural number linear combination of 32 and 50, which Theorem 4 tells us is the greatest common divisor of 32 and 50. Or, rather, they did up to the last step, where they finally "overshot"; the very last remainder was 0.

The most likely candidate for the greatest common divisor, then, is the last *nonzero* remainder, 2. Let us see why, regardless of the numbers a and b we started with, the last nonzero remainder in the chain of divisions we described above will be the greatest common divisor of a and b. First of all, we know that the last nonzero remainder is at least as large as (a, b), because Theorem 4 tells us that (a, b) is the smallest natural number linear combination of a and b.

What we will now do is show that the last nonzero remainder actually divides both a and b. That will mean that (a, b) is at least as large as the last nonzero remainder, so the two will be equal. Notice that in any division

$$d \overline{\smash{)}D} \overset{q r}{}$$

if the natural number c divides both the remainder and the divisor, then c also divides the dividend D. The reason is that, according to the definition of division,

$$D = q \cdot d + r.$$

Now, in the division chain we set up, the divisor and the dividend of each division step come from the remainder and the divisor of the step immediately preceding. Therefore, if the natural number c divides the divisor and the dividend of any one step, then c will divide the remainder and the divisor and so *also* the dividend of the step preceding. Therefore c will divide the remainder and the divisor of the step preceding that, so also the dividend of that step, and so on, until the conclusion is finally reached that c will divide both a and b.

Now, a particular example of such a number c is the last nonzero remainder. This number *is* the divisor and evenly divides the dividend of the very last division step. It follows, then, that this number divides remainder, divisor, and dividend of every step on up the chain of divisions, so it divides both a and b. (In our example, you can verify that the last nonzero remainder 2 is a divisor of every remainder, divisor, and dividend in each of the five division steps we went through.) This is what we needed to show to be certain that the last nonzero remainder would always be equal to (a, b).

The chain of divisions that we illustrated with the numbers 32 and 50 is what is called Euclid's algorithm. The algorithm is described as follows: Suppose a and b are two natural numbers, and that a is less than b. Then:

(1) Divide b by a.
(2) Divide a by the remainder of the above division.
(3) Divide the divisor of the preceding step by the remainder of the preceding step.
(4) Repeat step (3) until you obtain a remainder of 0.
(5) The last nonzero remainder is the greatest common divisor of a and b.

The notion of an *algorithm*, a process consisting of arithmetic and logical steps described in a fashion that anyone can follow them, is of great importance in our computerized age. All computer programs are algorithms. A computer must be told every single step it is to perform; nothing can be left to its imagination, because it has no imagination. (The word *algorithm* is a corruption of the name of an Arabian mathematician, Abu Ja'far Mohammed ibn Mûsâ al-Khowârizmî, who lived in the ninth century A.D. His textbook on arithmetic, *Hisâb al-jabr w'al-muqâbalah*, helped spread the use of the Hindu numeral system, which with modifications is the system we use today. The word *algebra* is a corruption of the second word in the title of his book.)

We will finish this section by proving a small but important theorem. A great many of the results in number theory are dependent on it. In fact, in the next chapter we will use it in several different places.

THEOREM 5. If a prime p divides the product ab of two natural numbers a and b, then it must divide a or b (or possibly both).

Proof. We will prove the theorem by showing that if p does not divide a, then it must divide b. Since p is a prime, it follows that if p does not divide a, then $(a, p) = 1$. By Theorem 4, then, there are integers r and s such that $1 = ra + sp$. Therefore,

$$b = b \cdot 1 = b(ra + sp)$$
$$= abr + pbs.$$

By assumption, p divides ab, so p divides abr. But p also divides pbs. Therefore p divides $abr + pbs = b$, as we were to show. This completes the proof.

EXERCISES

1. Find the following greatest common divisors:

    (1234, 6789)   (475, 323)   (693, 945).

2. List every number that appeared as divisor, dividend, and remainder in your use of Euclid's algorithm to find (693, 945) in Exercise 1. Verify that each of these numbers is a multiple of the greatest common divisor.

3. Find the greatest common divisor of the divisor and remainder of the following division:

    $$105 \overline{)4578}.$$

    Verify that the greatest common divisor is also a divisor of the dividend, 4578.

6. EUCLID'S ALGORITHM   39

4. Euclid's algorithm can at times be used for some really powerful sleuth work. Imagine a mathematical Sherlock Holmes presented with the following information: During January, a dealer's total sales on one model of television set were $1187. During February, his total sales on the same model were $1926. Sherlock concludes that the price per set is $107, that the dealer sold 11 sets in January and 18 sets in February. Tell how he arrived at his conclusions and what small assumptions he needed to make in order to reach the conclusions.

5. During August the friendly local door-to-door dictionary salesman had total sales of $6118, while in September his total sales were $5198. He sells only one model dictionary, for an integral number of dollars per copy, and the price per copy is less than $40, but greater than $5. What is the price per copy, and how many copies did he sell during each month?

6. A contractor is building a certain number of identical houses in a subdivision called Prestige Acres. They all are being built simultaneously. During March his total costs are $4473, and during April his total costs are $4375. Assuming that the cost per house per month is the same for each house and is an integral number of dollars, calculate how many houses are being built.

7. For the example of Euclid's algorithm in the text, write each remainder as a linear combination of 32 and 50.

8. The sequence of steps illustrated below with $a = 32$ and $b = 50$ yields a series of linear combinations of a and b which decrease in size.

$$32 \overline{)50} \quad \begin{array}{c} 1 \quad 18 \end{array}$$

$$18 \overline{)50} \quad \begin{array}{c} 2 \quad 14 \end{array}$$

$$14 \overline{)50} \quad \begin{array}{c} 3 \quad 8 \end{array}$$

$$8 \overline{)50} \quad \begin{array}{c} 6 \quad 2 \end{array}$$

$$2 \overline{)50} \quad \begin{array}{c} 25 \quad 0 \end{array}$$

Although in this case the last nonzero remainder turned out to be (a, b), one cannot hope that will always be true. Explain why not. Illustrate your explanation with an example in which the last nonzero remainder is not (a, b).

9. Prove that if the natural number c divides both the quotient and the remainder of a division, then c also divides the dividend.

10. Use Euclid's algorithm to find the greatest common divisor $(45, 96)$ of 45 and 96. By working back up through the divisions in the algorithm, write $(45, 96)$ as a linear combination of 45 and 96.

11. Suppose a small gear with 45 teeth is rolled around a large gear with 96 teeth. Suppose one of the teeth on the small gear is specially marked and in the beginning is touching the larger gear. (See the figure.) When the small gear is rolled around the large gear, the special tooth will come around and touch the large gear for the second time 45 teeth away from where it touched when the rolling started. Use your answer to Exercise 10 to show that on the 16th time the special tooth touches the large gear, it will touch just three teeth away from where it was touching when the rolling began. Will it ever touch just two teeth away or one tooth away from where it originally touched?

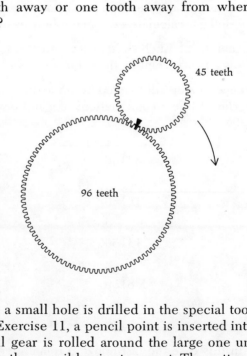

12. Suppose a small hole is drilled in the special tooth of the small gear in Exercise 11, a pencil point is inserted into the hole, and the small gear is rolled around the large one until the pattern drawn by the pencil begins to repeat. The pattern will have one

point for each time the special tooth touches the large gear. How many points will the pattern have? If you have a Spirograph (a toy manufactured by the Kenner Products Co.), you may be interested in drawing the pattern. In the Spirograph, the 96 teeth are on the inside of a ring, making the pattern a very pretty rosette.

CHAPTER THREE

Pythagorean Triples

1. The Theorem of Pythagoras

One of the most famous theorems in all geometry states that a triangle will have one of its angles a right angle if and only if the square of the longest side's length is equal to the sum of the squares of the lengths of the other two sides. (See Figure 1.) Tradition has always credited the Greek philosopher-mathematician Pythagoras (perhaps 570–500 B.C.) with being the first to discover a proof of this theorem. (Legend has it that he sacrificed 100 oxen to the gods in thanks for his good fortune.)

Regardless of Pythagoras' precise role in proving "his" theorem, it is now apparent that people of other lands — Babylonians, Egyptians, Indians — at least knew some special cases of the theorem. The Egyptians and the Indians used the special case of the 3, 4, 5 triangle in Figure 2 (made by stretching a knotted rope) to construct right angles. And, as we will see, more than 1000 years before Pythagoras, the inhabitants of Mesopotamia (whom we call Babylonians for simplicity) knew about the 3, 4, 5 triangle, the theorem of Pythagoras, and considerably more.

Think back to the time you first learned about the 3, 4, 5 triangle. Do you remember being at all surprised that by adding two perfect squares, 9 and 16, you could get another perfect square? Perfect squares are rather rare; between 1 and 100 there are only ten of them:

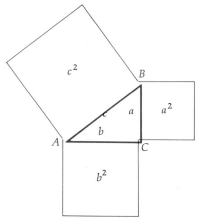

FIGURE 1. The theorem of Pythagoras says that the angle C of triangle ABC will be a right angle if and only if the area a^2 plus the area b^2 is equal to the area c^2.

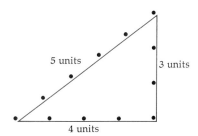

FIGURE 2. A loop of rope containing twelve equally spaced knots can be stretched to form a triangle having a perfect right angle, on account of the relation $3^2 + 4^2 = 5^2$.

1, 4, 9, 16, 25, 36, 49, 64, 81, and 100. Between 101 and 200 there are only four more: 121, 144, 169, and 196. The farther one goes out in the list of natural numbers, the more sparse become the perfect squares. We might expect, then, that it should be unusual for two perfect squares added together to yield another perfect square. In other words, we might regard a right triangle with sides of lengths a, b, and c all natural numbers as being a remarkable kind of triangle. Accordingly, we make the definition:

DEFINITION. A triple a, b, c of natural numbers which satisfies the equation

$$a^2 + b^2 = c^2$$

is called a *Pythagorean triple*.

44   CHAP. 3 PYTHAGOREAN TRIPLES

The triple 3, 4, 5 is the only Pythagorean triple in which all three numbers are less than 10. (This can be checked by a simple process of elimination.) Two other small examples of Pythagorean triples are 6, 8, 10 and 5, 12, 13. The triple 6, 8, 10 can be obtained from the triple 3, 4, 5 by multiplying each of the numbers 3, 4, and 5 by the number 2. On the other hand, the triple 5, 12, 13 cannot be obtained as a multiple of any smaller Pythagorean triple, and neither can the triple 3, 4, 5. We are thus led to distinguish a special kind of Pythagorean triple:

DEFINITION. A Pythagorean triple which is not a multiple of a smaller Pythagorean triple is called a *primitive* Pythagorean triple.

Notice that if we had a method for finding all primitive Pythagorean triples, then we would be able to obtain all Pythagorean triples as multiples of the primitive ones. The Greeks had a method for finding some Pythagorean triples, and they were able to prove that their method yielded all the primitive ones. (Their method also yields some triples which are not primitive, but it does not produce all the nonprimitive ones.) Until 1945, it was generally believed that the Greeks were the first people to know about that method.

However, in 1945, Otto Neugebauer and A. J. Sachs published a truly remarkable discovery concerning a list of numbers on an ancient Babylonian clay tablet. The tablet, familiarly known as "Plimpton 322" (that is, Item No. 322 in the Plimpton collection at Columbia University), had seemed much like any other Babylonian clay tablet from the period 1900 to 1600 B.C. until Neugebauer and Sachs took a very close look at it. It turned out that, except for four mistakes, two of the columns of numbers contained pairs of numbers from Pythagorean triples. (Furthermore, three of the four mistakes were readily explainable, being the clay tablet equivalents of typographical errors.)

The two columns of numbers, with the corrections, appear as columns a and c in Table 1. Now, what were the Babylonians doing with Pythagorean triples? The answer lies in the third important column on the tablet, which contains the numbers

$$\frac{c^2}{b^2} = \frac{c^2}{c^2 - a^2}$$

where b is the third number of the Pythagorean triple. In trigonometry, the ratio c/b is called the secant of the angle A.

1. THE THEOREM OF PYTHAGORAS

TABLE 1. Columns a, c, and $(c/b)^2$ are the columns found on Plimpton 322 (with corrections and restorations). The other three columns have been added to demonstrate what the Babylonian table is "all about."

a	b	c	$(c/b)^2$	Intended Angle	Square of Secant of Intended Angle
119	120	169	1.983	45°	2.000
3367	3456	4825	1.949	44°	1.933
4601	4800	6649	1.919	43°	1.869
12709	13500	18541	1.886	42°	1.811
65	72	97	1.815	41°	1.756
319	360	481	1.785	40°	1.704
2291	2700	3541	1.720	39°	1.656
799	960	1249	1.693	38°	1.610
481	600	769	1.643	37°	1.568
4961	6480	8161	1.586	36°	1.528
45	60	75	1.562	35°	1.490
1679	2400	2929	1.489	34°	1.455
161	240	289	1.450	33°	1.422
1771	2700	3229	1.430	32°	1.391
56	90	106	1.387	31°	1.361

Now the fifteen right triangles represented on Plimpton 322 have their angles A equal to approximately 45°, 44°, 43°, ···, 32°, and 31°. (See Figure 3.) Apparently Plimpton 322 was a trigonometry table—one of the world's oldest. We credit the Babylonians not only with being able to construct large Pythagorean triples (like 13500, 12709, 18541) but also with knowing, somehow, that all the triangles they were dealing with were right triangles. That is, they simply *must* have known the Pythagorean theorem, though perhaps they did not have a rigorous proof of it.

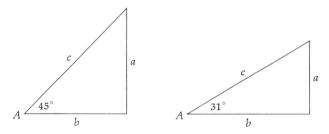

FIGURE 3. The Babylonian clay tablet Plimpton 322 lists values of a, c, and $(c/b)^2$ for fifteen different values of the angle A, beginning with A approximately 45 degrees and ending with A approximately 31 degrees.

How good a trigonometry table is Plimpton 322? By our standards today, it is not very accurate. In Table 1 you will find a column giving the squares of the secants of the angles 45°, 44°, 43°, $\cdots$, 32°, 31°, so that you can compare these values with the numbers c^2/b^2 calculated by the Babylonians. Probably, though, the numbers c^2/b^2 were sufficiently accurate for the Babylonians. To them, just as important as accuracy, was another condition which appears strange to us today. That condition is that the denominator b of the ratio c/b should divide some power of the number 60. All fifteen numbers in the b column of Table 1 satisfy this condition. As we will see once we ourselves have developed formulas for constructing Pythagorean triples, the formulas make it very easy to find triples satisfying the condition that b must divide some power of 60.

Can you remember when you first learned to carry the quotient of a division past the decimal point? You quickly noticed that certain divisions like

$$3\overline{)4}$$

produced unending decimals like

$$1.33333 \cdots.$$

If you felt uneasy about those never-ending decimals, your emotions were echoes of men's feelings over thousands of years. Whenever possible, the ancient mathematicians arranged things (with slight losses of accuracy if necessary) so that all quotients would terminate. For our decimal (base 10) number system, whenever a divisor evenly divides a power of 10, then regardless of what natural number the dividend may be, the quotient will be a terminating decimal. The Babylonians performed all their scientific computations in the sexagesimal (base 60) system. To assure that quotients in their sexagesimal system would always terminate, the Babylonian mathematicians took care to use only divisors which evenly divided some power of 60.

Today mathematicians readily accept any decimal, terminating or nonterminating, repeating or nonrepeating, as a perfectly legitimate number. The geometry of the Greeks started men on the road to acceptance, but it was the nineteenth century development of the foundations of calculus which enabled mathematicians to accept all decimals as numbers.

EXERCISES

1. What is the area of each of the four triangles in the figure below? Write down an expression for the area of the entire figure by

adding the areas of the four triangles to the area of the square in the center.

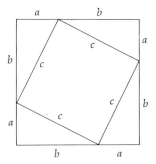

2. The area of the entire figure in Exercise 1 is also given by the expression

$$(a + b)^2 = a^2 + 2ab + b^2.$$

Equate this expression with the one you obtained in Exercise 1 and simplify, obtaining a proof of the theorem of Pythagoras. (This probably is not Pythagoras' original proof. There are some 370 different proofs of the theorem of Pythagoras which are presently known. One of them was invented by a president of the United States, James A. Garfield.)

3. By squaring the numbers a, b, and c, then comparing $a^2 + b^2$ with c^2, verify that the first four lines of Table 1 contain Pythagorean triples.

4. By elimination, prove that 3, 4, 5 is the only Pythagorean triple with all three numbers less than 10.

5. For each entry in column b of Table 1, find the smallest power of 60 which is divisible by that entry.

6. Verify that 125, 64, and 80 are all divisors of 10^6.

7. Perform each of the following divisions until the quotient terminates:

$$125 \overline{)3} \qquad 64 \overline{)58} \qquad 80 \overline{)179}.$$

8. Give an example of a division

$$a \overline{)b}$$

in which a divides no power of 10, yet the quotient terminates. [*Hint:* Choose a and b to have a greatest common divisor larger than 1.]

48   CHAP. 3 PYTHAGOREAN TRIPLES

9. Show that

$$\frac{169}{120} = 1 + \frac{24}{60} + \frac{30}{3600}$$

$$\frac{4825}{3456} = 1 + \frac{23}{60} + \frac{46}{3600} + \frac{2}{216000} + \frac{30}{12{,}960{,}000}.$$

(The two fractions on the left-hand sides of the equations are the ratios c/b from lines 1 and 2 of Table 1. The right-hand sides of the equations are the sexagesimal expansions of these ratios.)

10. Prove that any natural number multiple of a Pythagorean triple is again a Pythagorean triple.

11. Prove that if each number in a Pythagorean triple a, b, c is is divisible by the natural number d, then a/d, b/d, c/d is a Pythagorean triple.

12. Prove that if $(a, b) = 1$ and if the quotient of the division

$$a \overline{\smash{\big)}\,b}$$

terminates, then a must divide some power of 10. [*Hint:* Use Theorem 4 of the preceding chapter.]

13. (A small project.) In *The Exact Sciences in Antiquity* by Otto Neugebauer, Chapter 1, Section 12, you will find a description of how the Babylonians wrote numbers at the time Plimpton 322 was made. Write the columns a and c of our Table 1 in the Babylonian notation and compare them with the photograph of Plimpton 322 which you will find in the back of the same book.

2. Formulas for Constructing Pythagorean Triples

The following theorem gives the basic set of formulas for constructing certain Pythagorean triples (including, as we will later prove, all primitive ones). On the strength of the evidence, Plimpton 322, we feel quite sure that these formulas were known to the Babylonians of 1900 to 1600 B.C.

THEOREM 1. Let u and v be natural numbers, u larger than v. Let

$$a = u^2 - v^2,$$
$$b = 2uv,$$
$$c = u^2 + v^2.$$

Then a, b, c is a Pythagorean triple.

2. FORMULAS FOR CONSTRUCTING PYTHAGOREAN TRIPLES

Proof. We have
$$a^2 = (u^2 - v^2)^2 = u^4 - 2u^2v^2 + v^4$$
$$b^2 = 4u^2v^2.$$

Therefore,
$$a^2 + b^2 = u^4 + 2u^2v^2 + v^4$$
$$= (u^2 + v^2)^2$$
$$= c^2,$$

and the proof is complete.

The formulas of Theorem 1 enable us to produce many Pythagorean triples. In fact, letting $v = 1$ and letting u range over the values 2, 3, 4, 5, $\cdots$, we see that we obtain infinitely many values of a, b, and c. That is, the formulas of Theorem 1 generate infinitely many Pythagorean triples.

In Table 2 we see some of the smaller examples of Pythagorean triples generated by the formulas of Theorem 1. Three of them are primitive: 3, 4, 5; 5, 12, 13; and 15, 8, 17, and the other two are not primitive, because they are both multiples of the triple 3, 4, 5: 6, 8, 10 and 12, 16, 20. We are curious enough, then, to ask the question, Do the formulas of Theorem 1 produce *all* Pythagorean triples?

TABLE 2. Some examples of Pythagorean triples produced by the formulas of Theorem 1.

u	v	a	b	c	Primitive?
2	1	3	4	5	yes
3	1	8	6	10	no
4	1	15	8	17	yes
3	2	5	12	13	yes
4	2	12	16	20	no

The answer is that they do not; for example the triple 9, 12, 15 is not of the form given in Theorem 1. However, Theorem 2 below guarantees that they will at least produce all the primitive ones, and, of course, from the primitive ones we can obtain all the rest by multiplying by the various natural numbers. Theorems 1 and 2 constitute a full solution to the problem of finding all Pythagorean triples.

THEOREM 2. *If a, b, c is a primitive Pythagorean triple, then there are natural numbers u and v, u greater than v, such that either*
$$a = u^2 - v^2, \qquad b = 2uv, \qquad c = u^2 + v^2,$$
or
$$a = 2uv, \qquad b = u^2 - v^2, \qquad c = u^2 + v^2.$$

CHAP. 3 PYTHAGOREAN TRIPLES

Proving Theorem 2 takes some effort, but the theorem is such a remarkable one that it will be well worth the struggle. In order to get some feeling for where we stand, let us compare Theorems 1 and 2. Theorem 1 furnished us with many examples of Pythagorean triples. The triples were given by formulas, and in order to prove the theorem, all we had to do was take each triple and make sure that it satisfied the equation

$$a^2 + b^2 = c^2.$$

Now, proving that a given bunch of numbers satisfies an equation is a rather easy task. It consists in "plugging in" the numbers and making certain that the two sides of the equation are the same. In fact, that is exactly what we did to prove Theorem 1.

Proving Theorem 2, though, is the same as asking for all possible primitive solutions of the equation

$$a^2 + b^2 = c^2.$$

If a person is asked for just a few solutions to an equation, he can often produce them by guessing them and then verifying that they work. However, if he is asked for *all* solutions, then he can expect to put in some effort. He will start to manipulate the equation in some way or other, but each time he performs a step, he must be certain that he does not lose any solutions to the original equation.

What makes the problem especially challenging, of course, is the fact that we are interested only in natural number solutions. That means we cannot be so free-wheeling about the arithmetic we perform, because the operation of division is no respecter of natural numbers. In all, we can expect to have an interesting time working on Theorem 2. For clarity we will divide our work into two parts. In the next section we will prove two preliminary results. Once we have these two preliminaries, we will keep them in mind as in Section 4 we begin the search for all primitive solutions to

$$a^2 + b^2 = c^2.$$

Four different times in that search we will need the answer to some question, and each one of those four times we will not have to divert our attention from the search, because the question will have been answered in one of the theorems of Section 3.

The trick of parceling out the work is practiced by all mathematicians when their subject begins to bog down in details. You may sometimes have wondered how mathematicians can keep everything straight when they go so deeply into a subject. Part of keeping things straight is native ability, but the other part is organization like what you are about to see.

EXERCISES

1. Write down the nine Pythagorean triples obtained by letting $v = 1$ and $u = 2, 3, 4, 5, \cdots, 10$ in the formulas of Theorem 1.
2. For each of the triples you obtained in Exercise 1, find (a, b). In which five cases is $(a, b) = 1$? Notice that when (a, b) is equal to 1, the triple a, b, c must be a primitive triple. Are the other four triples primitive?
3. Show that the Pythagorean triple 9, 12, 15 is not of the form given in Theorem 1. [*Hint:* Find all natural numbers u and v, u greater than v, such that $2uv = 12$, and then show that $u^2 - v^2$ is never equal to 9 for any of these u and v.]
4. For each of the first five Pythagorean triples given in Table 1, find u and v. [*Hint:* First solve for u in terms of a and c.]
5. The natural number 25 is the number c in two different Pythagorean triples, one primitive and one not primitive. What are the two triples? [*Hint:* To find the primitive triple, find u and v such that $u^2 + v^2 = 25$, where u and v are natural numbers, u larger than v. Then use Theorem 1.]
6. Show that if u is even, then $a = u^2 - 1$ and $b = 2u$ have greatest common divisor equal to 1. [*Hint:* Show that (a, b) must be an odd number, so it therefore must divide u. Then show that a divisor of u which is also the divisor of $u^2 - 1$ must be equal to the number 1.] This result shows that infinitely many *primitive* Pythagorean triples are obtained by letting $v = 1$ and $u = 2, 4, 6, 8 \cdots$.

3. Preliminaries

In proving Theorem 2, we will be dealing only with primitive Pythagorean triples. We will now prove the three fairly simple facts illustrated in Figure 4 about primitive Pythagorean triples.

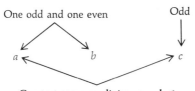

FIGURE 4. The three properties of a primitive Pythagorean triple proved in Theorem 3.

THEOREM 3. Let a, b, c be a primitive Pythagorean triple. Then

(i) $(a, c) = 1$.
(ii) c is odd.
(iii) Exactly one of a and b is even.

Proof. (i) If (a, c) is larger than 1, let p be any prime divisor of (a, c) larger than 1. Then a and c are both multiples of p, so a^2 and c^2 are both multiples of p, so

$$b^2 = c^2 - a^2$$

is a multiple of p. Because p divides $b^2 = b \cdot b$, by Theorem 5 of Chapter 2, p divides b. That is, a, b, and c are all multiples of the natural number p. We have shown that the triple a, b, c will not be primitive if (a, c) is greater than 1. Therefore, since we are assuming a, b, c is primitive, $(a, c) = 1$.

(ii) Since $a^2 + b^2 = c^2$, we have

$$(a + b)^2 = 2ab + c^2. \tag{1}$$

From Equation (1), we solve for ab to obtain the equation

$$ab = \frac{(a + b)^2 - c^2}{2}.$$

Now suppose c is even. Then c^2 is even, so $2ab + c^2$ is even, so from Equation (1) we see that $(a + b)^2$ is even. Therefore (by Theorem 5 of Chapter 2), $a + b$ is even. Therefore $(a + b)^2$ is a multiple of 4. Therefore the numerator of

$$\frac{(a + b)^2 - c^2}{2}$$

is a multiple of 4, so the fraction, which is equal to ab, is a multiple of 2. That is, ab is an even number. Therefore at least one of a or b is even. But $a + b$ was shown to be even, so then both a and b must be even. We have now shown that if c is even, all three of a, b, and c are even, so the triple a, b, c is not primitive. Therefore, since we are assuming a, b, c is primitive, c must be an odd number.

(iii) If a and b were both odd, then a^2 and b^2 would both be odd, so $c^2 = a^2 + b^2$ would be even, and so c would be even, contradicting what we proved in (ii). If a and b were both even, then a^2 and b^2 would both be even, so $c^2 = a^2 + b^2$ would be even, and c would be even, again contradicting what we proved in (ii).

This completes the proof of the theorem.

Let us add a word of comment on the proofs we have just gone through. Parts (i) and (iii) of the proof are of a very straightforward nature. There is just one simple idea underlying each, and if you remember the ideas, you will have no difficulty understanding and remembering those portions of the proof. Part (ii) falls into the category of "elusive" proofs. There is one key idea, and that is to show that if c is even, then so are a and b, so the triple cannot be primitive in that case. But, somehow, remembering how a and b were shown to be even is tricky for the average person.

You may well want to remember how the proof of part (ii) went, if only because you might be held responsible for it on a quiz. But, more important, you also may want to remember it in order to demonstrate to yourself that you really understood the proof when you read it. I have a special strategy to suggest in this case. *Without* looking back at the proof, but remembering the key idea, to show that a and b are both even, try recreating the proof on your own. You may not hit upon the same proof I gave, but if you work at it, you probably will work out one of at least half a dozen variations of the proof. Finding your own way, making up your own proof, can be a very ego-building activity, and I recommend it highly. The reason I think you have a good chance of success is that I worked out a number of variations and finally picked one of them for the proof to be put in this book. Any number of opening moves will lead to a successful proof, but in each case there will of course be some work to be done.

We will now prove a second result. This theorem can be used in contexts far removed from Pythagorean triples, because it makes no mention of them whatever.

THEOREM 4. Let x and y be natural numbers such that $(x, y) = 1$. If there is a natural number z such that

$$xy = z^2$$

(that is, if xy is a perfect square), then both x and y are perfect squares.

Proof. If the theorem is false, pick from all products $xy = z^2$ for which it is false, a product for which z is as small as possible. That is, we are assuming that $(x, y) = 1$, that $xy = z^2$, and that at least one of x and y is not a perfect square. We will use these conditions plus the minimality of z to derive a contradiction. The conclusion finally will be that there is no minimal z for which the theorem is false, and so the theorem must always be true.

By changing notation if necessary, we can assume that x (at least) is not a perfect square.

If the square p^2 of some prime p greater than 1 were to divide x, then we would have

$$\frac{x}{p^2} \cdot y = \left(\frac{z}{p}\right)^2.$$

Because we picked z to be the smallest number for which the theorem is false, the theorem must be true for the natural numbers x/p^2 and y. That is, x/p^2 and y must be perfect squares. But then

$$x = p^2 \frac{x}{p^2}$$

would also be a perfect square, contrary to our choice of x. The conclusion is that x is never divisible by the square of a prime larger than 1.

Because x is not a perfect square, x must be larger than 1. Therefore x must be divisible by at least one prime p larger than 1. Therefore $z^2 = xy$ must be divisible by that prime p. Since p divides z^2, then by Theorem 5 of Chapter 2, p divides z. Therefore z^2 is divisible by p^2, so the product xy is divisible by p^2. Now we have already shown that p^2 does not divide x, but p does divide x. Therefore, since

$$\frac{x}{p} y$$

is divisible by p and x/p is not, then by Theorem 5 of Chapter 2, y must be divisible by p. But then both x and y are divisible by p, so (x, y) is greater than 1, which contradicts one of the conditions placed on x and y in the statement of the theorem.

Therefore, there can be no smallest z for which the theorem is false. This concludes the proof.

EXERCISES

1. Show that in *any* Pythagorean triple at least one of a and b must be even.

2. Use the example 48, 14, 50 of a (nonprimitive) Pythagorean triple to illustrate the three parts to the proof of Theorem 3.

3. For any primitive Pythagorean triple, not only is it true that $(a, c) = 1$, but also that $(a, b) = 1$ and $(b, c) = 1$. Prove these additional two results. [*Hint:* You can use almost the same argument as in Theorem 3.]

4. Is it possible for a Pythagorean triple to satisfy both conclusions (ii) and (iii) of Theorem 3 and yet not be primitive? Give a proof for your answer.

5. Use the example $63 \cdot 7 = 21^2$ to illustrate the proof of Theorem 4 down to the sentence, "The conclusion is that x is never divisible by the square of a prime larger than 1." Then use the example $7 \cdot 63 = 21^2$ to illustrate the rest of the proof.

6. Find a different proof of part (ii) of Theorem 3.

4. Proof

In this section we will put together a proof of Theorem 2, drawing upon all that we proved in the preceding section. For handy reference, let us restate the theorem here.

THEOREM 2. If a, b, c is a primitive Pythagorean triple, then there are natural numbers u and v, u greater than v, such that either

$$a = u^2 - v^2, \qquad b = 2uv, \qquad c = u^2 + v^2,$$

or

$$a = 2uv, \qquad b = u^2 - v^2, \qquad c = u^2 + v^2.$$

Proof. By part (iii) of Theorem 3, exactly one of the numbers a and b must be even. If necessary, let us switch our notation so that b is the even number. We will show that in this case we are led to the first of the two lines of equations above.

First of all, we rewrite the equation

$$a^2 + b^2 = c^2$$

as

$$\begin{aligned} b^2 &= c^2 - a^2 \\ &= (c+a)(c-a). \end{aligned}$$

Now a is odd, and by part (ii) of Theorem 3, c is also odd. Therefore both $c + a$ and $c - a$ are even. Let x and y be the natural numbers such that

$$\begin{aligned} c + a &= 2x, \\ c - a &= 2y. \end{aligned}$$

Consider the greatest common divisor (x, y) of x and y. Both $x + y$ and $x - y$ are divisible by (x, y), so we find that (x, y) divides

$$x + y = \tfrac{1}{2}(c + a) + \tfrac{1}{2}(c - a) = c$$

and also

$$x - y = \tfrac{1}{2}(c + a) - \tfrac{1}{2}(c - a) = a.$$

But by part (i) of Theorem 3, the only common divisor of a and c is 1. Therefore $(x, y) = 1$.

Now

$$\begin{aligned} xy &= \tfrac{1}{2}(c + a)\tfrac{1}{2}(c - a) \\ &= \tfrac{1}{4}(c^2 - a^2) \\ &= \tfrac{1}{4}b^2 \\ &= \left(\frac{b}{2}\right)^2. \end{aligned}$$

Since b is even, $b/2$ is a natural number, so x and y are two numbers whose product is a perfect square and such that $(x, y) = 1$. Therefore, by Theorem 4 we conclude that both x and y are perfect squares. Let u and v be the natural numbers such that

$$\begin{aligned} x &= u^2, \\ y &= v^2. \end{aligned}$$

We saw above that $x - y = a$, a number greater than zero. Therefore x must be larger than y, so u must be larger than v.

And now, using some of the equations we derived earlier in the proof, we find

$$\begin{aligned} a &= x - y = u^2 - v^2, \\ c &= x + y = u^2 + v^2. \end{aligned}$$

Finally, from the equation

$$xy = \left(\frac{b}{2}\right)^2,$$

we have

$$\begin{aligned} b &= 2\sqrt{xy} \\ &= 2\sqrt{u^2 v^2} \\ &= 2uv. \end{aligned}$$

We have now completed the proof of Theorem 2.

Let us end this section by returning to the Babylonians' use of the formulas for Pythagorean triples. You will recall we mentioned that the Babylonians constructed Pythagorean triples a, b, c with the property that the number b divided some power of 60. You can go back to Table 1 and check quickly that every one of the fifteen entries in the b column has this property. (In fact, every one of those numbers divides $60^4 = 12,960,000$.) How did the Babylonians guarantee this would happen? Well, since the formula for b is simply $b = 2uv$, all they had to do was make certain that each of the two numbers u and v was a divisor of some power of 60. Then b also would be a divisor of some (possibly higher) power of 60.

As to what method they used to pick triples whose triangles had angles close to 45°, 44°, 43°, $\cdots$, 31°, we have no idea. That part of the process has not been found on or deduced from any of the considerable number of clay tablets uncovered and studied so far. It is estimated that there have been a half million tablets found to date (although only a small portion of these have been identified as mathematical tables), and there are many more as yet unexcavated, so perhaps someday we will know the full story.

EXERCISES

1. Pick any one of the primitive Pythagorean triples from Table 1, and use it to illustrate the proof of Theorem 2. (All triples in Table 1 are primitive, except for those in lines 11 and 15.)

2. It was stated in Section 2 that the triple 9, 12, 15 is not obtainable from the formulas of Theorems 1 and 2. Push the triple 9, 12, 15 through the proof of Theorem 2 to see where the failures of the proof occur.

3. Find the generating numbers u and v for the fourteen triples in Table 1 which have generating numbers u and v. For which triple can you not find the numbers u and v? Of what primitive triple is it a multiple?

4. If u and v generate a primitive Pythagorean triple, show that $(u, v) = 1$. [*Hint:* Show that (u, v) divides a, b, and c.]

5. If u and v generate a primitive Pythagorean triple, show that exactly one of u and v must be even. [*Hint:* If they are both odd, show that c must be even, contrary to Theorem 3.]

6. (Hard exercise.) If u and v are natural numbers such that u is greater than v, $(u, v) = 1$, and exactly one of u and v is even, show that the Pythagorean triple they generate is primitive.

7. Prove that Theorem 2 holds for nonprimitive Pythagorean triples a, b, c in which $(a, c) = 2$. [*Hint:* Use most of the proof of Theorem 2, but insert a few more steps to show that $(x, y) = 1$.]

8. Use Exercise 7 to show that any Pythagorean triple a, b, c in which (a, c) is a power of 2 can be obtained from the formulas of Theorem 2. [*Hint:* Divide a, b, c by a power of 4 to produce a triple in which $(a, c) = 1$, $(a, c) = 2$, $(b, c) = 1$, or $(b, c) = 2$.]

5. Fermat's Last Problem

We will close this chapter by telling the remarkable sequel to the story of Pythagorean triples. You will recall that we wondered a little about the ability of natural numbers to mesh so precisely that the equation

$$a^2 + b^2 = c^2$$

could be true. Yet we found a way of producing infinitely many triples a, b, c for which the equation is satisfied.

Pierre de Fermat wondered if it could ever be true that

$$a^n + b^n = c^n \tag{2}$$

for a, b, c, and n natural numbers with n larger than 2. In fact, he did more than wonder; he was firmly convinced that (2) never held under those conditions. He wrote in the margin of a book he owned, ". . . I have found a wonderful proof of this, but the margin is too small to contain it." The note was found in the book after Fermat's death, and that is the only allusion to the problem in all of his writings. A proof of what he believed to be true has never been found, although many capable mathematicians have worked on the problem.

However, various people have found Fermat's conjecture to be true in a number of special cases. The biggest step ever taken toward finding a proof was made by the German mathematician E. Kummer (1810–1893). He found quite a few values of n for which he was able to show that Equation (2) would never hold for natural numbers a, b, and c. Kummer worked on Fermat's problem by developing a whole new branch of mathematics, called algebraic number theory, which he used in his study of Equation (2).

At one point he actually believed that he had completely solved the problem, but it turned out he had made one small mistake. When the mistake was gotten around, he was able to show that for certain values of n, (2) could never be satisfied by natural numbers. It turned out later that Kummer's algebraic number theory could be applied to many other problems throughout mathematics. Algebraic number theory is today one of the well-established subjects in modern mathematics and has itself led to the development of still other fields of mathematics. Fermat's problem, by the amount of mathematics it has inspired, has proved to be one of the greatest problems in the history of mathematics. Whether it eventually is found to be true or false, its importance will always lie more outside itself than within.

In 1908 Paul Wolfskehl bequeathed a prize of 100,000 marks to be awarded to the first person who completely solved Fermat's problem. The Academy of Science at Göttingen (Germany) was named to administer the prize. From the four corners of the world, a flood of attempted proofs descended upon the Academy. Each one was studied by graduate students, who located the first mistake and then sent a reply to the author. The Weimar Republic's inflation reduced the prize to almost nothing, and in turn the "last problem" correspondence died down.

It has been shown by various people that Fermat's conjecture is true for all natural numbers n less than 25,000. Therefore, if Equation (2) is ever satisfied for n greater than 2, the numbers a^n, b^n, and c^n involved in it must be truly enormous.

CHAPTER FOUR

Pioneering Results in Topology

1. Introduction

The preceding two chapters of this book were devoted to number theory, a subject primarily concerned with *quantitative* relationships. In this chapter and the next we will study a subject which is primarily concerned with spatial relations. If you studied plane geometry in high school, you have already seen one example of a mathematical subject which deals with space. The objects studied in geometry were things like triangles, line segments, angles, and circles, and the principal relationships between them were congruence and similarity. Two figures are congruent if, as in Figure 1, one can be placed right on top of the other; they are similar if, as in Figure 2, the one can be placed right on top of the other after one has been enlarged or reduced an appropriate amount.

FIGURE 1. Two congruent triangles.

In plane geometry, then, one is often concerned with moving objects around and even enlarging or reducing them, but ordinarily one does not consider deforming them: bending lines, straightening

1. INTRODUCTION

FIGURE 2. Two similar triangles.

arcs, changing angles, stretching figures. However, there are properties of figures that remain unchanged under all sorts of bending and stretching deformations and that are very interesting to study. One simple example is the following property of any circle: It divides the plane in which it is drawn into two areas, the inside of the circle and the outside. If the circle is deformed into a triangle, the triangle still has the property of dividing the plane into two areas. (See Figure 3.)

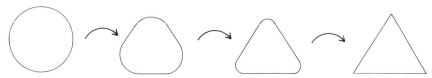

FIGURE 3. A circle being deformed to a triangle.

The name given to the subject which studies these very fundamental kinds of properties is *topology*. The men who first did some work in the subject seem to have been René Descartes (1596–1650) and Leonhard Euler. This chapter will be devoted to their work. They seem to have been a bit like the Norse discoverers of America; at the times they lived the mathematical world was not ready to take advantage of their new ideas. The subject did not really begin to grow until after 1850, and what mathematicians know as topology today is really a twentieth century invention.

Although topology is a comparatively recent subject, students usually find it easier to grasp (at least at the beginning level) than they do number theory, which seems to have been developed as far back as 4000 years ago. Perhaps the reason is that in topology the examples that one uses to aid his thinking are pictures (which are fun to draw), while the corresponding examples in number theory are usually tables of numbers.

EXERCISES

1. From the topological point of view, the six figures below consist of three pairs of equivalent figures. (Two figures are topologically equivalent if one can be deformed to the other by any

means short of cutting apart or pasting together.) Tell which figures are equivalent to which. Which figures divide the plane into exactly two areas?

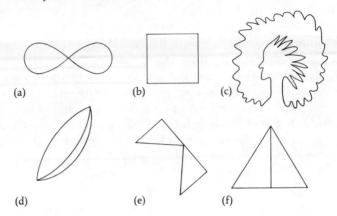

2. From the topological point of view, some of the 26 capital letters A, B, C, D, E, F, G, H, I, J, K, L, M, N, O, P, Q, R, S, T, U, V, W, X, Y, and Z are equivalent, because it is possible to convert certain letters into certain others by straightening, bending, stretching, shrinking (but not cutting or pasting). For example, C, I, and L are all equivalent, and D and O are equivalent. Divide the 26 letters into sets of equivalent letters by placing in the same set any two letters that can be deformed into each other.

3. The number of sides to a surface is a property that is unchanged under any deformations short of cutting and pasting. An example of a surface with two sides is an ordinary sheet of paper. A thin strip of the sheet of paper with its ends pasted together without twisting, to form a band, also has exactly two sides. But the same strip of paper with its ends pasted together after a half-twist forms a band with only one side. This single-sided surface is called a *Möbius band* (after August Ferdinand Möbius (1790–1868)). Describe the difference between a Möbius band and an ordinary band in terms of an insect crawling on each kind of surface. (For a spectacular illustration of a Möbius band, complete with nine ants crawling on it, see Reproduction 40 of *The Graphic Work of M. C. Escher,* Meredith Press, New York, 1967.)

4. If you cut an ordinary paper band lengthwise down the middle, it will become two bands each half as wide as the original. What will happen if you make the same cut in a Möbius band? (If

you have *any* doubt as to what will happen, try it; the experience of seeing "it" happen right before your eyes is unforgettable.)

5. Draw the following figure without lifting pencil from paper and without retracing any line: [*Hint:* If you start at either A or B you will have little or no trouble.]

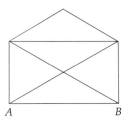

2. The Königsberg Bridge Problem

The city of Königsberg (now called Kaliningrad) is located at the confluence of two branches of the river Pregel. There is an island in the river, and the city is split all told into the four parts A, B, C, and D of Figure 4. At the time the Königsberg bridge problem was proposed, there were seven bridges linking the various parts of the city, as shown in the figure.

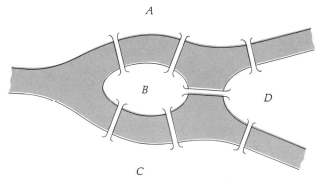

FIGURE 4. The seven bridges of Königsberg (1736 A.D.).

The problem is a curious one: Can one take a walking (no swimming, please) journey through Königsberg so as to cross each bridge once and only once? You might be interested to try a few times. One way to settle the problem, of course, would be to try all possible routes. Either they would all fail, or there would be at least one which was successful. Either way, the problem would be settled. (There are only a finite, but large, number of routes conceivable, since each different route must cross the seven bridges in some different order, and there are only 7! = 5040 ways in which to order the seven bridges.)

However, it seems that no one settled the problem by this method. (One might call such an approach a "method of exhaustion," for two different reasons. Little wonder that no one tried it.) Leonhard Euler settled the problem by proving a general theorem which showed that certain problems of the type to which the bridge problem belongs cannot be solved. Then he simply observed that the configuration of the Königsberg bridges causes that problem to be one which his theorem says cannot be solved. Euler's theorem is usually stated in terms of topological objects called *graphs*, rather than islands, rivers, and bridges.

DEFINITION. A *graph* is any finite set of points together with certain line segments or arcs joining these points. The points are called the *vertices* of the graph, and the line segments or arcs are called the *edges* of the graph.

DEFINITION. A graph is said to be *connected* if it is possible to travel from any vertex to any other vertex by moving along edges. (See Figure 5.)

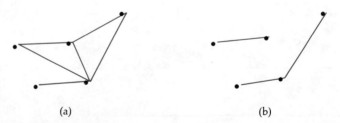

(a) (b)

FIGURE 5. (a) A connected graph (5 vertices, 6 edges). (b) A nonconnected graph (5 vertices, 3 edges).

The graph that corresponds to the bridges of Königsberg is shown in Figure 6. The four land areas of the city have been replaced by vertices (labeled the same way), and the bridges connecting them have been replaced by edges. The Königsberg bridge problem is equivalent to the question: Can one trace this graph without lifting pencil from paper and without tracing any edge more than one time? The answer is given by Euler's theorem.

2. THE KÖNIGSBERG BRIDGE PROBLEM

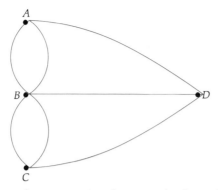

FIGURE 6. The graph representing the seven bridges of Königsberg.

THEOREM 1 (Euler). Consider any graph that is connected. Such a graph can be traced without lifting pencil from paper and without tracing any edge more than once if and only if at most two vertices have an odd number of edges emanating from them.

We will prove the "only if" part of the statement here. That will be sufficient to settle the Königsberg bridge problem, because in Figure 6 all four vertices have odd numbers of edges emanating from them. Since we are going to show that the tracing can occur only if there are at most two vertices with that property, it will follow that the graph *cannot* be traced, and so the proposed journey through Königsberg is *impossible*.

Proof. Call a vertex with an odd number of lines emanating from it an "odd vertex." We will suppose we have a tracing of the prescribed kind in a graph with more than two odd vertices and derive a contradiction from this supposition. The tracing may or may not have *begun* at an odd vertex. Regardless, since there are at least three odd vertices in the graph, there are at least two odd vertices at which the tracing did *not* start. Call these two vertices A and B.

We will concentrate on the progress of the tracing at vertex A. It has to pass through A at various times until at last all the edges emanating from A are used up. When it comes to A along one edge, it must leave along another (since edges may not be traced more than once), and then those two edges can never be traced again. Thus, each time the tracing passes through A, it uses up a pair of edges until at last there is only one edge left. When it finally traces this edge, it comes to A and then cannot leave, since all edges from A have been used up. Therefore the tracing must end at A.

But to prove this, all we used was the information that A is an odd vertex at which the tracing did not begin. Since B is also such an odd vertex, the same argument shows that the tracing must end at B as well. But A and B are different vertices, so the tracing must end in two different places at the same time. This is the contradiction we were looking for. The proof is complete.

EXERCISES

1. Which of the following graphs can you trace without lifting pencil from paper and without tracing any edge more than once?

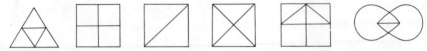

2. For the graphs in Exercise 1 which can be traced, perform a tracing. [*Hint:* The idea behind the proof of Theorem 1 will tell you where to start.]

3. After Euler's time an eighth bridge was added, spanning the river downstream from the island in Königsberg. Show by an example path that this eighth bridge makes it possible to take a walking journey which crosses each of the eight bridges exactly once.

4. In Sherman Stein's *Mathematics: The Man-made Universe*, Euler's graph-tracing problem is called the problem of the "thrifty highway inspector." Thinking of the edges of a graph as highways, explain why that name is an appropriate one for Euler's problem.

5. A salesman for the Fly-by-Night Company wishes a route through a number of cities such that he *never* passes through the same city twice. Imagine the graphs of Exercise 1 to be maps, where each edge represents a highway and each vertex a city. Through each graph, draw a route on which the salesman will pass through each city (vertex) once and only once.

6. In which of the graphs below are there Fly-by-Night salesmen's routes passing through all the vertices? Draw the routes where they exist. Explain, as best you can, why the routes do not exist for each case in which the routes do not exist.

7. Show that, if a graph contains at least one odd vertex, then it contains at least two odd vertices; that if it contains at least three odd vertices, then it contains at least four odd vertices.

8. Prove that no graph can contain an odd number of odd vertices. Use this fact to show that the number of people who have shaken hands an odd number of times is even. (Other acts involving pairs of people can be substituted for handshaking if you so desire.)

9. Suppose you are asked, without lifting pencil from paper, to trace a connected graph tracing each edge exactly *twice*. Is this always possible regardless of the number of odd vertices?

10. Suppose you are asked to trace a connected graph as usual *except* that you may if you wish pick any one edge to be traced twice. Can you trace the graph regardless of the number of odd vertices if you are given this option?

11. The Fly-by-Night salesman problem (also called the problem of drawing a Hamiltonian circuit in a graph) has been around for 100 years without a satisfactory answer. That is, there is no general result like Theorem 1 for this second kind of graph-tracing problem. Show that Theorem 1 (apparently) has nothing to do with this problem, by constructing three graphs in which the salesman's circuit can be drawn — one with no odd vertices, one with two odd vertices, and one with four odd vertices.

12. Continuing Exercise 11, construct three connected graphs in which the salesman's circuit cannot be drawn — one with no odd vertices, one with two odd vertices, and one with four odd vertices.

3. Duality

Euler's theorem, although it is stated in terms of tracing a graph, can be used to solve other problems if one can devise a way of converting them to graph tracing problems. We settled the Königsberg bridge problem by making such a conversion. There are many such problems that can be converted by passing to what is called a *dual* configuration.

For instance, a puzzle that is more frequently seen today than the Königsberg bridges (perhaps because it is easier to draw) is the problem of drawing a continuous curving line through every one of the edges of the graph shown in Figure 7. The conditions are that you may not pass through the same edge twice, and you may not pass through any vertex. Although Figure 7 is a graph, Euler's theorem does not apply immediately because, of course, the problem is not one of tracing Figure 7.

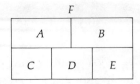

FIGURE 7. The problem is to draw a continuous curving line which passes through each edge exactly once (and through no vertices).

However, it is easy to find a graph-tracing problem that is equivalent to this problem. The puzzle hinges on passing in and out of the various areas, just as in the Königsberg bridge problem. And to convert the problem to one of graph tracing, we do just as we did with the areas of the city—we represent each of the six areas of Figure 7 by a vertex. (The sixth area is the outside of the figure.) Then for each edge of Figure 7 lying between two areas, we draw an edge joining the corresponding vertices. The result is shown in Figure 8.

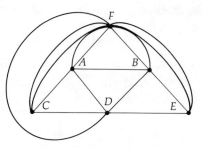

FIGURE 8. The graph dual to that of Figure 7.

The graph of Figure 8 is called the *dual* to the graph of Figure 7. From the way we have constructed the dual, the original problem is equivalent to tracing the dual graph without lifting pencil from paper and without tracing any edge more than once.

Thus, to settle the problem, all we need to do is count how many odd vertices there are in the dual graph. We find that there are four, namely A, B, D, and F. Therefore the dual graph cannot be traced, and so the original problem cannot be solved. Notice that one does not actually need to draw the dual graph to apply Euler's theorem. All that is necessary to know is the number of odd vertices in the dual, and this can be found by looking at each area in the original figure to see whether it is bounded by an even or an odd number of edges.

One of the purposes of this section was simply to show you a crude example of the phenomenon known as duality. Dualities of various kinds can be found in practically all branches of mathematics, but they are especially common in topology. Dualities are so many and so varied that they defy being pinned down by a definition of "what duality is." However, in this one example you have seen how useful the concept of duality can be, for by it we translated a problem in which we were interested but had done no work on into another problem about which we know the whole story.

In certain branches of mathematics, as soon as a theorem is proved, one can use duality to translate that theorem to another statement. That second statement can then be considered proved, because one can write down a proof for it simply by writing down the dual of every statement appearing in the proof of the original theorem. Thus, as a person proves a series of theorems, he obtains free of charge a dual theory, consisting of the series of theorems which are the duals of the theorems he is proving.

EXERCISES

1. For which of the following figures can you draw a continuous curving line which passes through each line segment exactly one time? In those cases where it is possible to draw the line, give an example of such a line.

2. Which of the figures in Exercise 1 can you draw without lifting pencil from paper and without retracing any line segment? Show how to draw those which can be drawn.

3. Suppose a house has doorways as in the figure following. Is it possible to walk through each doorway exactly once? If there were no back door to the house, would it be possible to walk through each doorway exactly once? If such a walking journey is possible, draw it.

70 CHAP. 4 PIONEERING RESULTS IN TOPOLOGY

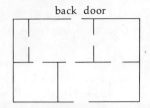

4. Draw the first floor plan of your own house, showing all doorways. Is it possible to walk through each doorway on the first floor of your house exactly once?

5. Given a graph drawn on a sheet of paper, there are a total of four cases with respect to the problems of drawing the graph and drawing the continuous curving line:

	Draw the Graph	Draw the Line
1	Possible	Possible
2	Possible	Impossible
3	Impossible	Possible
4	Impossible	Impossible

Show by giving examples that each of these four cases can arise. [*Suggestion:* Use your previous work.]

6. Find the dual of each of the following graphs.

7. Find the dual of the dual of each of the graphs in Exercise 6.

8. In the statement of the problem of drawing a continuous curving line through the edges of a graph, there was a condition that the line must never pass through a vertex. What condition does this translate to in the problem of tracing the dual graph?

4. An Application of Graphs to Business Management

In recent years graphs have been used to solve a variety of important problems in industry. Sometimes these uses have been very sophisticated, making use of considerable theory, while others have

4. AN APPLICATION OF GRAPHS TO BUSINESS MANAGEMENT

been beautifully simple. They all point to the worth of knowing something about graphs. One rather simple application of graphs is known as the PERT chart. It is a device used to keep tabs on the progress of projects that must meet important deadlines.

The PERT chart was first used in connection with the truly immense project of developing the missile firing Polaris submarine fleet. It is estimated that use of the PERT technique (plus the efforts of a very dedicated team of managers) may have shaved two years from the total development time of Polaris. The word PERT is an acronym for "Program Evaluation and Review Technique," but this author has never been impressed with this particular series of words. Therefore, we will make use only of the abbreviation PERT and not of the words themselves.

Put yourself in the shoes of a project manager in some aircraft company, and suppose you have just been given the project of producing a flying model of a new kind of airplane. The plans are all in, and the components are all available; all you have to do is put the pieces together. Let us suppose that the construction of the airplane consists of eight major tasks, as follows:

1. Building the frame.
2. Installing the control surfaces.
3. Installing the engine.
4. Installing the fuel tanks.
5. Installing the control equipment.
6. Installing fuel lines.
7. Installing control lines.
8. Attaching the skin to the frame.

Some of these tasks can be worked on simultaneously, while some cannot be started until certain others are finished. Suppose that the priorities of the tasks are as given in Table 1.

TABLE 1. Priorities that must be observed in performing the eight tasks involved in building the airplane.

Task	Requires Completion of Tasks
1	
2	1
3	1
4	1
5	1
6	3, 4
7	2, 3, 5, 6
8	1, 2, 3, 4, 6, 7

Two important questions that should pop into the mind of you, the project manager, as soon as you see the table, are:

(1) How can I best picture the relationships given in the table?
(2) How can I calculate how long it will take to complete construction of the airplane?

Of course, a table the size of Table 1 is not difficult to comprehend as is, but imagine the difficulty of seeing the interrelationships listed in a table for a project involving several hundred tasks. Modern projects often involve that many tasks. (In fact, the construction of an airplane may well involve that many tasks; our example is a deliberately simplified one.)

The two questions we have asked were so important in the Polaris project that a special contract was let to a group of management consultants to find the best way of answering them. The consultants found that the best way was to construct a special kind of graph in which all of the tasks were represented by edges. Such a graph is now called a PERT chart.

The edges in the PERT chart which represent tasks will be called *task edges*. Generally, the task edges are drawn horizontally or else as nearly horizontally as possible. The reason is that we wish to think of the left end of the task edge as representing the beginning of the task, and of the right end as representing the end of the task, as is illustrated in Figure 9.

```
BEGIN  •———— Task 6 ————•  END
TASK 6   (Installing fuel lines)  TASK 6
```

FIGURE 9. Task edges are drawn horizontally, or as nearly horizontally as possible. The left end represents the beginning of the task, and the right end represents the completion of the task.

There is one other kind of edge in a PERT chart, called a *constraint edge*. Each constraint edge is used as an indicator that a certain task cannot be begun until some other task is ended. This is done by drawing the constraint edge from the beginning of the one task back to the end of the task whose completion is necessary for beginning the new task. For example, because Table 1 says that Task 6 cannot be started without Task 4 being ended, we have the constraint edge joining the two task edges in Figure 10.

We are now ready to produce a PERT chart representing Table 1. First of all, we draw (horizontally) the 8 task edges roughly in the order in which we think the tasks will have to be performed. See Figure 11. The next step is to join them by constraint edges, according to Table 1. This causes the tangle we see in Figure 12.

4. AN APPLICATION OF GRAPHS TO BUSINESS MANAGEMENT

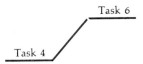

FIGURE 10. To represent the fact that starting Task 6 is dependent upon completing Task 4, we draw a constraint edge from the right end of the Task 4 edge to the left end of the Task 6 edge.

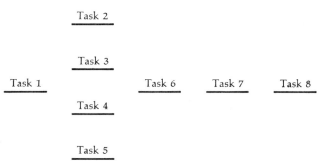

FIGURE 11. The first step in producing a PERT chart is to draw the task edges in approximately the order in which the tasks will have to be done.

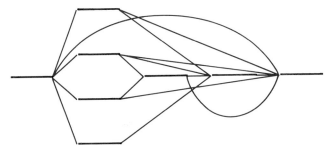

FIGURE 12. The next step in producing a PERT chart is drawing in all the constraint edges. The result is usually a huge tangle, which fortunately can be simplified.

Happily, Figure 12 is unnecessarily complicated, because some of the constraints portrayed in it are redundant. For example, the constraint edge joining the end of Task 2 to the beginning of Task 8 is not needed. It says that we cannot begin Task 8 until Task 2 is ended. But, by looking at Figure 12, we see that this restriction is already implied by the constraints which say that Task 8 depends upon Task 7, which in turn depends upon Task 2. Therefore, it is safe to erase the constraint edge between Tasks 2 and 8. Five more constraint edges can be eliminated in the same way, and thus we find that the much simpler Figure 13 also portrays the priorities listed in Table 1.

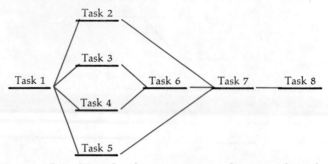

FIGURE 13. After the redundant constraint edges are cleared away, the resulting PERT chart gives quite a clear picture of the priorities listed in Table 1.

To this day, the kind of graph shown in Figure 13 is the best way known of representing priorities among tasks. Such a graph is known as a PERT chart. It is the answer to our first question. To show that it really does make priority relations easy to study, we will show how it can be used to calculate the answer to our second question: How long will the project take?

In order to answer this question, we naturally require information as to the length of time required to complete each task. In practice, such information is always obtained as time estimates from the people in charge of the various tasks. Suppose the times are those given in Table 2.

TABLE 2. The number of days required to complete each task.

Task	Number of Days Required
1	20
2	3
3	7
4	4
5	10
6	2
7	5
8	15

The first step is to write in these times on the corresponding task edges in the PERT chart, as in Figure 14. If you like, you can begin thinking of each constraint edge as representing a task which requires 0 time to be accomplished; in that way, every edge will have a number associated with it.

4. AN APPLICATION OF GRAPHS TO BUSINESS MANAGEMENT

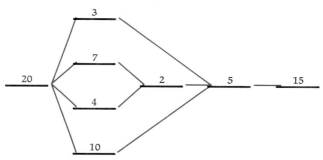

FIGURE 14. In order to determine the minimum amount of time in which a project can be completed, the first step is to write above each task edge the estimated time which that task will require.

The next step is to work through the PERT chart from right to left, assigning to each vertex the *minimum* amount of time before completion of the project at which the Begin or End represented by that vertex can occur. The three rules for making these assignments are:

(1) A right-most vertex receives the number 0, since it represents completion of the entire project. (In our example, then, End Task 8 is labeled with the number 0.)

(2) To label a Begin vertex, add the amount of time the task requires to the number with which the End vertex for that task has already been labeled.

(3) To label an End vertex, assign to it the largest number found on the various Begin vertices which are joined to the End vertex by constraint edges.

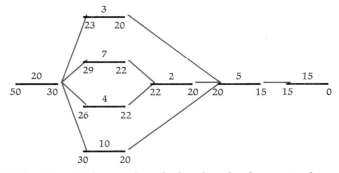

FIGURE 15. Then below each end of each task edge, write the minimum amount of time before completion of the project at which the event represented by that end can occur. This is done by working from right to left through the PERT chart, as described in the text. The largest number at the far left of the chart, 50 in this case, will be the minimum time from beginning to end of the project.

You will see, as you carry out these rules in any chart, that you will naturally do the labeling from right to left. Figure 15 shows our example of a PERT chart with all its vertices labeled.

Now, how do we find the minimum length of time in which the project can be completed? We look at all the labels of Begin vertices from which no constraint edges emanate. In our example there is only one such vertex, Begin Task 1, but it is possible that there may be more. The *maximum* number in those labels will be the *minimum* amount of time in which the project can be completed. Thus, the construction of our hypothetical airplane can take place in a minimum time of 50 days.

We have now answered the two questions that we posed at the beginning of this section. There are two other topics in PERT, which we will mention only briefly. The PERT chart represents the entire project as a graph all of whose edges must be traversed from left to right before the project is completed. Some path through the PERT chart must take a longest time to traverse. (That longest time in our example would be 50 days.) Such a path is called a *critical path*, because a delay in any one of the tasks along that path will delay the completion of the entire project. The project manager will make use of another computation to locate that path, and he will pay special attention to seeing that tasks along the critical path are completed within the estimated times. Our example, Figure 15, is simple enough that we do not need a computation to find the path. It is the bottommost route through the graph.

The other topic in PERT is the *updating* of the chart. It may happen that some task off the critical path will be severely delayed, resulting in some different path assuming the role of the critical path. The manager usually updates his chart weekly to see if such a thing has happened and if he will be able to finish the project on time.

We mentioned that PERT was first developed in connection with the Polaris submarine project. That project was so complicated that the master PERT chart stretched for 50 feet along the wall of a large room. The computations for such a large project, of course, are not done by hand. There are special computer programs available for doing the necessary PERT computations, all the way to updating. The success of the technique has been so impressive that most large government contracts are limited to contractors who use PERT.

EXERCISES

1. Suppose the 9 courses in the table below are the minimum requirements for a B.S. in physics in a certain college. Write a

4. AN APPLICATION OF GRAPHS TO BUSINESS MANAGEMENT 77

PERT chart for the project consisting of obtaining the B.S. in physics. Calculate the minimum amount of time in which the project can be completed, and locate the critical path.

Course	Number of Semesters	Requires Completion of Courses
1. Basic Physics	3	
2. Calculus	2	
3. Differential Equations	1	2
4. Mechanics	2	2, 3
5. Electromagnetics	1	1, 2, 3
6. Thermodynamics	1	1, 2
7. Relativity	1	4
8. Quantum Mechanics	1	4, 5
9. Senior Research	1	4, 5, 6, 7, 8

2. Repeat Exercise 1 for whatever subject you are majoring in or considering majoring in. (Pick out definite courses to fill in all of your major electives. Also, be sure to include any courses outside the major which are prerequisites for the major courses — like the calculus and differential equations courses in Exercise 1.)

3. Show by an example that a PERT chart can have two different paths, both of which are critical.

4. Make up your own PERT example with 10 or more named tasks. Carry out the operations of simplifying the PERT chart, calculating the minimum time to completion, and locating the critical path. (For an example, you might take the project of building an automobile, cooking Thanksgiving dinner, baking a cake, producing a play, or whatever strikes your fancy.)

5. The *slack time* of a task is the additional amount of time that can be allotted to completing that task without delaying the completion of the project (assuming all other tasks are completed in the exact amount of time estimated for their completion). For instance, there is slack time of 1 day in Task 3 of the airplane example in the text. Compute the slack times for the other seven tasks of the airplane example.

6. Compute the slack times for the various courses in the project of obtaining a B.S. in physics, in Exercise 1.

7. Give a general rule for computing slack times in any PERT chart. [*Suggestion:* Work from left to right through the chart.]
8. Fill in the blank: Every task along a critical path has _____ slack time. (If one is going to the trouble of computing slack times anyway, simply because they are useful to know, it is very easy then to use them in locating the critical path.)

5. The Descartes-Euler Formula for Polyhedra

A polyhedron is defined to be any solid figure all of whose bounding faces are polygons. The surface of a polyhedron is thus made up of three kinds of geometrical objects: polygons or *faces*, line segments or *edges*, and points or *vertices*. Figure 16 shows the faces, edges, and vertices of a tetrahedron (four-sided polyhedron).

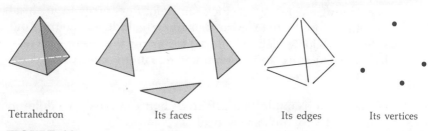

Tetrahedron   Its faces   Its edges   Its vertices

FIGURE 16.

Any polyhedron whose surface can be deformed by bending and stretching until it becomes a sphere is called *simple*. Generally, a nonsimple polyhedron will fail to be simple for one of two reasons. Either it has tunnels through it, or it is composed of two or more polyhedra joined along vertices or edges. Figure 17 shows each of these two cases alongside the simple tetrahedron.

Let us agree to denote the number of faces of a polyhedron by F, the number of edges by E, and the number of vertices by V. For any simple polyhedron, there is a remarkable formula which relates $F, E,$ and V.

THEOREM 2 (Descartes-Euler). For any simple polyhedron,

$$F - E + V = 2. \tag{1}$$

5. THE DESCARTES-EULER FORMULA FOR POLYHEDRA

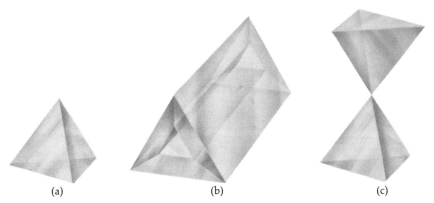

FIGURE 17. One simple polyhedron and two nonsimple polyhedra. (a) Simple. (b) Nonsimple (bored). (c) Nonsimple (waspish).

One might suppose that this formula, or at least some special case of it, was discovered by the Greek mathematicians. They would certainly have had a high regard for it, inasmuch as it has both a number-theoretic and a geometric flavor. Above all, it is spectacularly beautiful in its simplicity. However, they apparently knew nothing of it. It seems to have been discovered first by René Descartes and then later independently by Leonhard Euler.

You will see from the proof we are about to give that Theorem 2 is topological in nature. We will reduce Theorem 2 almost immediately to a statement about a graph drawn in a plane, and thereafter the word polyhedron will not even be mentioned.

Proof of Theorem 2. Remove any one face from the polyhedron. Because of the simplicity of the polyhedron, the remaining surface can be deformed until it becomes flat, a connected graph lying in a plane. (Figure 18 illustrates this process for a cube.) The removal of the one face of the polyhedron has given the graph one face fewer than the polyhedron had, but the number of edges and the number of vertices have both remained unchanged. Therefore, to prove Theorem 2, it will be enough to prove:

THEOREM 3. Consider any connected plane graph each of whose edges bounds a face and each of whose faces is deformable to a polygon. If F, E, and V are respectively the number of faces, edges, and vertices of the graph, then

$$F - E + V = 1. \qquad (2)$$

Proof. Consider any face of the graph with n edges, n greater than 3. If an edge is added joining two vertices which are themselves joined to a common vertex by edges of the face, then the face is split

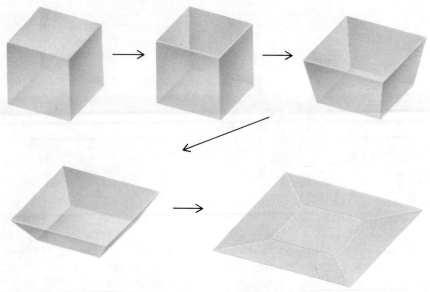

FIGURE 18. Deforming a cube less one face to a plane graph.

into two smaller faces, one bounded by 3 edges and one bounded by $n-1$ edges. See Figure 19. By this addition, the number of faces increases by 1, and the number of edges increases by 1, but the number of vertices is unchanged. We thus obtain a new graph with the same value of $F - E + V$ as the original.

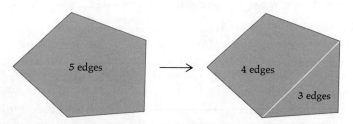

FIGURE 19. Adding an edge in the process of triangulation.

As long as there is a face with more than 3 edges, we continue to add edges in this fashion until at last we obtain a graph each of whose faces is bounded by exactly 3 edges. We will call these faces triangles, although it is possible their edges may be curved. At no stage does the value of $F - E + V$ change, so the value of $F - E + V$ for the "triangulated" graph is the same as for the original graph. Therefore it will be sufficient to prove that $F - E + V = 1$ for the special case of a connected graph all of whose faces are triangles. We will do this by

5. THE DESCARTES-EULER FORMULA FOR POLYHEDRA

building such a graph from a single triangle, adding one triangle at a time, and at each step making certain that the value of $F - E + V$ is unchanged.

Pick any triangle within the graph. The value of $F - E + V$ for the triangle alone is $1 - 3 + 3 = 1$, as it is supposed to be. To this triangle we will add the triangles of the graph, one by one, such that at each step the triangle added has at least one vertex in common with what is already present. Furthermore, at each step we will avoid enclosing any empty space within the figure obtained by that step.

(We can avoid enclosing empty spaces as follows: If at any time the triangle we choose to add will cause a space to be enclosed, *instead* choose a triangle within the empty space and again touching the figure already present. This may still cause an empty space to be enclosed, but this space will be made up of fewer faces of the graph. Thus, this process of rechoosing can be continued until at last no empty space will be enclosed. See Figure 20 for an illustration of this rather long-winded set of instructions.)

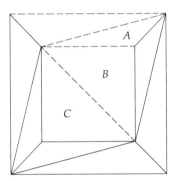

FIGURE 20. The triangulated graph obtained from a cube. The solid-edged faces have already been added. Adding triangle A will enclose an empty space. Instead, add one of the triangles within that empty space which touch the figure already present—either B or C. If B is chosen, an empty space is still enclosed, so pick a triangle within that empty space—C.

Since the original graph is connected, if we continue the process of adding triangles according to the above rules, we will ultimately obtain the entire graph. What we now have left to do is show that the value of $F - E + V$ never changes if we observe the rules we have agreed to for adding each new triangle. There are several cases to be considered, depending upon how many edges are added when the new triangle is adjoined.

Case 1. One edge is added. See Figure 21. In this case, two of the edges are already present and consequently all three vertices are present. Thus, the total number of faces increases by 1, the total number of edges increases by 1, but the total number of vertices stays the same. Thus the value of $F - E + V$ stays the same.

FIGURE 21. Illustration of Case 1; one edge and one face are added.

Case 2. Two edges are added. See Figure 22. In this case, one edge is already present, and so at least the two vertices joined by this edge are present. If the third vertex were present also, addition of the triangle would enclose an empty space between the figure already present and the new triangle. Since we are avoiding this, the third vertex is not present. Thus, we are adding one vertex, two edges, and one face. Consequently, the value of $F - E + V$ stays the same.

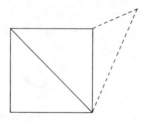

FIGURE 22. Illustration of Case 2; one vertex, two edges, and one face are added.

Case 3. Three edges are added. See Figure 23. In this case there is one vertex already present in the figure (that is one of the rules for adding triangles), but the other two are not, since if one or both were, we would be enclosing an empty space. Thus we are adding one face, three edges, and two vertices, and so again $F - E + V$ stays the same.

"Case 4." No edges are added. This case cannot occur, since it would consist of filling in an empty space, and we are adding triangles in such a way that empty spaces never arise.

5. THE DESCARTES-EULER FORMULA FOR POLYHEDRA

Therefore the entire graph can be built in steps starting from a single triangle, in such a way that at each step the value of $F - E + V$ is preserved. Therefore the value of $F - E + V$ for the entire graph is the same as for the single triangle, namely $F - E + V = 1$. This completes the proof of Theorems 2 and 3.

FIGURE 23. Illustration of Case 3; two vertices, three edges, and one face are added.

EXERCISES

1. Calculate $F - E + V$ for each of the graphs below by counting the number of faces, edges, and vertices in each case.

2. The tetrahedron (Figure 17(a)) had four faces and four vertices. Can you think of several other polyhedra with $V = F$? [*Hint:* Imagine yourself in Egypt.]

3. Draw two different plane graphs, both of which have $V = F$. In each case, count the number of edges in the graph. Is either number of edges an even number?

4. Prove that any plane graph for which $V = F$ must have an odd number of edges. [*Hint:* Replace V with F in Equation (2).] You can make up a nasty puzzle along this line. Ask a friend to draw a graph with 8 edges having the same number of vertices as faces.

5. What is the value of $F - E + V$ for the graph below? For what two reasons does Theorem 3 not apply to this graph?

6. What is the first step in the proof of Theorem 3 which does not work for the graph of Exercise 5?

7. Suppose a graph with polygonal faces is made up of three connected graphs not connected to each other. What is the value of $F - E + V$? Do you see a method for calculating the number of connected pieces of which a plane graph with polygonal faces is made?

8. Find the value of $F - E + V$ for the nonsimple polyhedron of Figure 17(b).

9. Join two polyhedra of the type of Figure 17(b) by placing them side-by-side along a face, to make a polyhedron with two parallel holes through it. Find $F - E + V$ for this polyhedron.

10. Make up a polyhedron with one hole other than the one of Figure 17(b) and calculate $F - E + V$ for it. From this and the previous two exercises, do you think you might know how the value of $F - E + V$ is related to the number of holes in a polyhedron?

11. Prove that the assumptions of Theorem 3 can be relaxed a bit so as to allow two-sided faces in addition to faces which are deformable to polygons.

12. Carry out in detail the steps in the proof of Theorem 2 (including the proof of Theorem 3) for a cube as an example. Add triangles in each of the three possible ways, to illustrate Cases 1, 2, and 3.

13. A rather amusing problem of which you may have heard is called the utilities problem. Three houses are to be connected to three utilities, water, gas, and electricity, so that the lines do not cross each other. The figure below shows a frustrated attempt at making the connections. Use Theorem 3 to show this problem cannot be solved. [*Hint:* Make the assumption that you can solve the problem, and then you will have many nice

facts to lead to a contradiction. To get you started, if the problem can be solved, the solution will be a graph having 6 vertices and 9 edges.] There seems to be no practical value to this problem other than its worth as a brain-teaser for those not in on the secret.

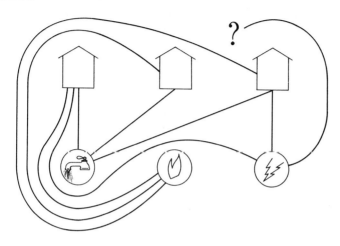

6. Regular Polyhedra

It would be a shame to have proved such a beautiful result as the Descartes-Euler formula without giving at least one important application of it. So we will give one now.

DEFINITION. A *regular polyhedron* is any polyhedron whose faces are all congruent to the same regular polygon and whose vertices all have the same number of edges emanating from them.

It turns out that there are only five kinds of simple regular polyhedra, having 4, 6, 8, 12, and 20 faces, respectively. Figure 24 shows the five, each labeled by its usual name. In four out of the five cases, the name is derived from the Greek word for the number of faces.

The Greeks knew of these five polyhedra; in fact, they set such great store by them that the polyhedra are sometimes called by the name of one of their greatest philosophers: Platonic solids. The disciples of Pythagoras, noted for their flights of fancy over the relationships between mathematics, religion, music, natural science, and so on, associated four of the five Platonic solids with the four "elements":

tetrahedron = fire
cube = earth
octahedron = air
icosahedron = water.

(The Pythagoreans came fairly early in the history of Greek mathematics, around 500 B.C., when the dodecahedron was not yet known. For that reason, the four-and-four correspondence was especially appealing. Imagine their embarrassment when one of their own members discovered the regular dodecahedron.)

The regular tetrahedron, the cube, and the regular octahedron all occur as crystal forms in nature. The regular dodecahedron and icosahedron cannot occur as crystal forms, but there are dodecahedral and icosahedral crystals which are very close to being regular polyhedra.

The Greeks had a somewhat long-winded proof that the five polyhedra of Figure 24 were the only simple regular polyhedra. We will give a modern proof, using the Descartes-Euler formula, of a theorem similar to but not quite the same as the one the Greeks knew.

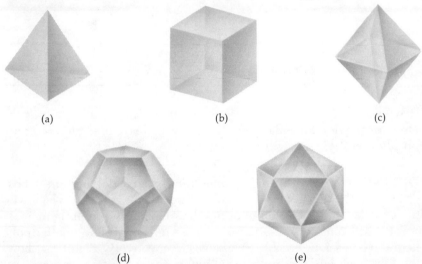

FIGURE 24. The five regular polyhedra. (a) Tetrahedron (4 faces). (b) Cube (6 faces). (c) Octahedron (8 faces). (d) Dodecahedron (12 faces). (e) Icosahedron (20 faces).

THEOREM 4. Consider a simple polyhedron with the following properties:

(i) Every face is bounded by the same number r of edges.
(ii) Every vertex has the same number n of edges emanating from it.

Then that polyhedron has the same values of F, E, V, r, and n as one of the five regular polyhedra of Figure 24.

Notice that this theorem does not assume regularity of the polyhedron, but that any regular polyhedron certainly satisfies the conditions (i) and (ii). Properties (i) and (ii) are not changed under deformation of a polyhedron and so are of a topological nature. One should expect, then, that the conclusion of the theorem would also be topological – and so it is. The conclusion says, in essence, that the polyhedron in question is some deformed version of those shown in Figure 24. It is not difficult to prove, using this conclusion plus a little geometry, that if the polyhedron is regular, then it is similar to one of the five of Figure 24.

Proof of Theorem 4. The first step is to write the two quantities V and F in terms of E, n, and r.

The quantity E is the number of edges in the polyhedron, so $2E$ is the number of ends of the edges. By assumption (ii), these $2E$ ends are clumped together, n at a time, to form the vertices, since each vertex has exactly n edges emanating from it. Therefore the polyhedron has $V = 2E/n$ vertices.

Now consider the possibility of writing the number of edges in terms of the number of faces. Each face is bounded by exactly r edges, according to assumption (i), but then each edge bounds exactly two different faces. Therefore the total number of edges is $E = Fr/2$. Solving for F, we have $F = 2E/r$.

Now we enter these two expressions for F and V in the Descartes-Euler formula:

$$\frac{2E}{r} - E + \frac{2E}{n} = 2. \tag{3}$$

With a little manipulation, we can convert this equation to one which is easy to study. First, divide by $2E$:

$$\frac{1}{r} - \frac{1}{2} + \frac{1}{n} = \frac{1}{E}. \tag{4}$$

Then transpose the $\frac{1}{2}$:

$$\frac{1}{r} + \frac{1}{n} = \frac{1}{2} + \frac{1}{E}. \tag{5}$$

Equation (5) is very restrictive on the values that r, n, and E can take. The basic idea is that for large values of r and n, $1/r + 1/n$ is a small fraction, while Equation (5) can only be satisfied when $1/r + 1/n$ is greater than $\frac{1}{2}$. Therefore we should expect only a few small values of r and n to be solutions of (5).

88 CHAP. 4 PIONEERING RESULTS IN TOPOLOGY

Now both r and n are at least equal to 3, since they are the number of edges bounding a polygon or emanating from a vertex. On the other hand, if both r and n were at least equal to 4, we would have $1/r + 1/n$ less than or equal to $\frac{1}{2}$, so Equation (5) could not be satisfied. Therefore, at least one of r and n must be equal to 3. The remainder of the proof will be divided into two cases, depending upon whether $r = 3$ or $n = 3$.

Case $r = 3$. Substitute the value 3 for r in Equation (5):

$$\frac{1}{n} + \frac{1}{3} = \frac{1}{2} + \frac{1}{E}. \tag{6}$$

Then rewrite Equation (6) as:

$$\frac{1}{E} = \frac{1}{n} - \frac{1}{6}. \tag{7}$$

For values of n greater than or equal to 6, $1/n - 1/6$ is not a positive number, so for those values Equation (7) cannot be satisfied. Therefore the only possible values for n are 3, 4, and 5. Let us see what each one of these values leads to.

Subcase $n = 3$. Substituting this value of n in Equation (7), we have

$$\frac{1}{E} = \frac{1}{3} - \frac{1}{6} = \frac{1}{6},$$

so $E = 6$. Then $F = 2E/r = 12/3 = 4$, and $V = 2E/n = 12/3 = 4$. These values and those of n and r are those of the tetrahedron, Figure 24(a).

Subcase $n = 4$. Substituting this value of n in Equation (7), we have

$$\frac{1}{E} = \frac{1}{4} - \frac{1}{6} = \frac{1}{12},$$

so $E = 12$. Then $F = 2E/r = 24/3 = 8$, and $V = 2E/n = 24/4 = 6$. These values and those of n and r are those of the octahedron, Figure 24(c).

Subcase $n = 5$. Substituting this value of n in Equation (7), we have

$$\frac{1}{E} = \frac{1}{5} - \frac{1}{6} = \frac{1}{30},$$

so $E = 30$. Then $F = 2E/r = 60/3 = 20$, and $V = 2E/n = 60/5 = 12$. These values and those of n and r are those of the icosahedron, Figure 24(e). This finishes the case $r = 3$.

6. REGULAR POLYHEDRA

Case n = 3. Substitute the value 3 for n in Equation (5):

$$\frac{1}{3} + \frac{1}{r} = \frac{1}{2} + \frac{1}{E}. \tag{8}$$

Then rewrite Equation (8) as:

$$\frac{1}{E} = \frac{1}{r} - \frac{1}{6}. \tag{9}$$

For values of n greater than or equal to 6, $1/r - 1/6$ is not a positive number, so for these values Equation (9) cannot be satisfied. Therefore the only possible values for r are 3, 4, and 5. Let us see what each of these values leads to.

Subcase r = 3. We have already analyzed this case of $n = r = 3$ and found the values of E, F, V, r, and n to be those of the tetrahedron, Figure 24(a).

Subcase r = 4. Substituting this value of r in Equation (9), we have

$$\frac{1}{E} = \frac{1}{4} - \frac{1}{6} = \frac{1}{12},$$

so $E = 12$. Then $F = 2E/r = 24/4 = 6$, and $V = 2E/n = 24/3 = 8$. These values and those of n and r are those of the cube, Figure 18(b).

Subcase r = 5. Substituting this value of r in Equation (9), we have

$$\frac{1}{E} = \frac{1}{5} - \frac{1}{6} = \frac{1}{30},$$

so $E = 30$. Then $F = 2E/r = 60/5 = 12$, and $V = 2E/n = 60/3 = 20$. These values and those of n and r are those of the dodecahedron, Figure 24(d). This completes the proof.

EXERCISES

1. Carry through the steps in the proof of Theorem 4 for the cube as a particular example of a polyhedron satisfying the assumptions of the theorem.

2. The edges and vertices of any polyhedron form a graph. For which of the five regular polyhedra can the graph consisting of the edges and vertices be traced in the fashion prescribed in Theorem 1? Give an example of such a tracing for those cases in which it can be done.

3. Compare the values of F, E, and V for the cube and the octahedron. Do you notice anything that might be worth discussing? [*Hint:* Recall the contents of Section 3.]

4. If the cube and octahedron are duals of each other, and if the dodecahedron and icosahedron are duals of each other, what is the dual of the tetrahedron?

5. Find all natural number solutions to the equation

$$\frac{1}{r} + \frac{1}{n} = \frac{1}{2}.$$

(There are three pairs satisfying it.)

6. The equation of Exercise 5 may be thought of (loosely) as arising from Equation (5) by letting E be infinite. Draw portions of the plane figures which correspond to each of the three solution pairs r, n of Exercise 5. [*Suggestion:* First draw one polygon face of r sides, then add edges at each vertex until there are n edges at each vertex, then draw more polygon faces using these edges, and so on.]

7. Prove that in any simple polyhedron E is greater than or equal to $3V/2$ and greater than or equal to $3F/2$. [*Hint:* Use the facts that every face is bounded by at least 3 edges and that every vertex has at least 3 edges emanating from it.]

8. Use Exercise 7 to prove that any simple polyhedron with exactly 4 faces must have at least 6 edges.

9. Use Exercise 7 together with Theorem 2 to prove that any simple polyhedron with exactly 4 faces must have at most 4 vertices.

10. Combine Exercises 8 and 9 to show that any simple polyhedron with exactly 4 faces must have:

    (1) Exactly 6 edges.
    (2) Exactly 4 vertices.
    (3) Exactly 3 edges emanating from each vertex.
    (4) Exactly 3 edges bounding each face.

    This result says that any simple tetrahedron is topologically equivalent to a regular tetrahedron.

11. It turns out that there are two topologically different kinds of simple polyhedra which have exactly 5 faces. Sketch an example of each kind. [*Hint:* One kind has exactly 3 faces which are bounded by 4 edges.]

12. Prove that in any simple polyhedron the number V of vertices is less than or equal to $2F - 4$. [*Hint:* The proof of Exercise 9 is a special case of the proof which works here.]

13. Prove that in any simple polyhedron the number F of faces is less than or equal to $2V - 4$. [*Hint:* The proof of this fact can be obtained by reversing the roles of F and V in the proof of Exercise 12.]

14. (A dual to Exercise 11.) There are two topologically different kinds of simple polyhedra which have exactly 5 vertices. Sketch an example of each kind.

15. Is it possible for all the faces of a simple polyhedron to be congruent, yet the polyhedron not be topologically equivalent to a regular polyhedron? Prove your answer. [*Hint:* What happens if you butt one regular tetrahedron up against another one?]

16. (Challenge exercise.) Prove that the two polyhedra you sketched in Exercise 11 are the only two topologically different simple polyhedra which have exactly 5 faces. [*Hint:* Derive your inspiration from Exercises 7, 8, 9, and 10.] You may be interested to know that matters quickly get out of hand as the number of faces increases; there are 7 different kinds of simple polyhedra with 6 faces, 34 kinds with 7 faces, and 257 kinds with 8 faces!

CHAPTER FIVE

Map Coloring

1. Introduction

In this chapter we will focus our attention on a famous problem in topology. The problem is now over 120 years old. Forty years after it was first proposed, it was almost completely solved. The emphasis should be placed on the word "almost," because 80 years more have passed, and no one has yet succeeded in finding the complete solution.

Around 1850 a British student named Francis Guthrie was coloring a map of England. All of a sudden, he began to wonder, if a person wished to color a map so that regions which border on each other are colored contrastingly, what is the greatest number of colors he would ever need? At first glance, it might look as though the number must be quite large, because maps can get extremely complicated—in the mind of a mathematician even more so than in real life. Yet Guthrie was unable to find *any* example of a map which required more than four colors. There is the stuff of which a good mathematical question is made. The first glance says one thing, while a little experimentation says just the opposite. Guthrie was intrigued. He tried to find a proof that no more than four colors are ever necessary, but his efforts were unsuccessful.

He passed the question on to his brother, who in turn asked his mathematics professor, Augustus de Morgan, if he knew the answer. This took place in 1852. De Morgan did not know the answer, and he

1. INTRODUCTION

in turn tried to interest some of his friends in studying the problem. No one showed much interest until, in 1878, Arthur Cayley (1821–1895) announced at a meeting that he had tried and was unable to solve Guthrie's problem. Cayley was the most famous English mathematician of the nineteenth century, so his announcement created quite a stir. The next year a lawyer named A. B. Kempe published a false proof that no more than 4 colors were necessary. For ten years no one found the flaw in Kempe's proof, and so during that time the problem was believed to be solved. Finally, in 1890 P. J. Heawood found the error in the proof. Heawood revised Kempe's argument and found that the method of proof did show that no more than 5 colors are necessary.

Now, there are maps for which 4 colors are required. Figure 1 shows a map having 4 regions, each region bordering on the other three. For this map, then, 4 colors are needed. The Kempe-Heawood proof says that no map, however complex, will need more than 5 colors. But *is* there a complicated map which requires 5 colors, or might someone eventually find a proof that 4 colors will always be enough? Eighty more years have passed, and the question is still unanswered.

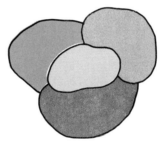

FIGURE 1. A map with four regions, each bordering on the other three. This map requires four colors in order that each region may contrast with all regions bordering on it.

We will begin a gradual buildup to a proof of the Kempe-Heawood 5-color theorem. As a first step, let us reach agreement concerning a few properties of maps. For us, all maps will consist of only a finite number of regions. Each separate region will be called a "country." For example, we will consider Pakistan to be two countries, because it is composed of two separate regions. When we speak of coloring a map, we will always mean coloring it in the way Francis Guthrie prescribed, so that no two countries with a common boundary are given the same color. (Two countries that touch only at isolated points are not regarded as having a common boundary.) Until further notice we will assume all our maps lie in a plane. Later we will take up the matter of maps on the surface of a sphere (and on the surface of a doughnut, even).

The key to the proof of the Kempe-Heawood theorem is a clever application of the Descartes-Euler formula for connected plane graphs, Theorem 3 of the preceding chapter. But what graph do we apply the Descartes-Euler formula to? We apply it to the map itself. We think of the countries as being faces of a plane graph, the boundary segments as being the edges, and the points where the boundary segments come together as being the vertices. For example, if we think very hard of Figure 1 as being a graph, in our mind's eye we might see it turning into something like Figure 2.

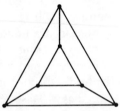

FIGURE 2. Concentrating on the boundary segments and vertices of Figure 1 as the edges and vertices of a graph, we might imagine Figure 1 straightening out into this shape.

EXERCISES

1. Color each of the maps below with as few colors as possible.

2. How many colors are needed to color contrastingly the faces of a tetrahedron so that any pair of faces with an edge in common are colored differently?

3. Remove one face and deform the rest of the tetrahedron, cube, and octahedron to plane graphs (having respectively 3, 5, and 7 faces). Color the faces of each of these graphs with as few colors as possible.

4. Tell why the map of the continental United States (excluding Alaska and Hawaii) cannot be colored with fewer than 4 colors. [*Hint:* Count how many states border on Nevada.]

5. Are there other states besides Nevada from which you can readily show that the map of the 48 states requires 4 colors?

6. Color a map of the 48 states with 4 colors in such a way that if two states have a common boundary they are colored differently.

7. We mentioned in this section that we are considering each separate region on a map to be a different country. Which one of the 48 states, then, is actually two "countries"? (Forget about islands in the Atlantic or Pacific.)

8. If one requires that two countries be given different colors even when they touch only at an isolated point, how many colors would be necessary to color all maps drawn in a plane?

9. Repeat Exercise 2 for the two topologically different kinds of 5-faced polyhedra. (See Chapter 4, Section 6, Exercise 11.)

10. Repeat Exercise 3 for the regular dodecahedron and icosahedron.

2. Some Properties of a Map When Viewed as a Graph

What kind of country should we allow to appear in a map? On first thought, we might just say: Any old blob, as long as it is in one piece. However, by saying "blob," we are limiting ourselves unnecessarily. It is possible that a country would not be a solid blob, but rather (as in Figure 3) like a ring, or a set of brass knuckles, or a slice of Swiss cheese—having holes in which are located one or more other countries. For example, the country of Italy completely surrounds the Vatican, which is an independent country complete with laws, mail, diplomats, army, and so on. We will, then, prove the 5-color theorem for maps having not merely blob-shaped countries but also Swiss-cheese-shaped countries.

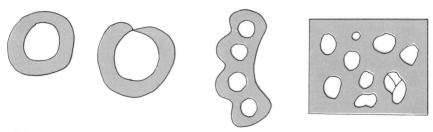

FIGURE 3. A map may have countries which have one or more holes and countries within those holes. We will first prove the 5-color theorem for maps without such countries. The 5-color theorem for the more general kinds of maps is easy to prove once the one special case is established.

However, we will be able to defer our attention to the cheese until later. We will first prove the 5-color theorem for maps containing only blob countries. Then, when we have completed that proof, we will show that the more general situation can be reduced to the special case by means of slits which open the holes in the Swiss cheese. Therefore, we will first consider only countries containing no holes. When such a country is viewed as the face of a graph, we see that either it is deformable to a polygon or else it is a "monogon" or a "digon," as in Figure 4.

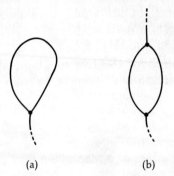

(a)   (b)

FIGURE 4. (a) Monogon. (b) Digon. Once we restrict ourselves to the case where no country has a hole in it, every country must be deformable to a polygon or else be of one of the two shapes pictured here.

If a monogon occurs, it cannot border on another country, because if it did, that country would surround it and hence be a country with a hole in it. Therefore, all monogon countries can be given any colors we choose, either before or after the rest of the map is colored. *Therefore, we can restrict our attention to maps containing only polygon and digon countries.*

Now let us take a look at the vertices of a map. The vertices occur only where boundary segments come together. It is impossible to have just two boundary segments come together at a point, because if just two did they would split up the same two countries, and so they would be just one boundary segment. *Therefore, when we view our map as a graph, we see that every vertex has at least three edges (or boundary segments) emanating from it.*

What would happen if our map does not form a connected graph? Well, it would break up into two or more connected pieces, and the problem of coloring the map with at most 5 colors could be solved by coloring each of the connected pieces with at most 5 colors. *Therefore, to prove the 5-color theorem, it will be enough to prove it for maps which are connected when viewed as graphs.*

2. SOME PROPERTIES OF A MAP

Finally, suppose a map has some digon (2-boundary) countries. By inserting a tiny circular country at a vertex of each digon country as in Figure 5 (small enough so as not to touch or cover any other vertex), we can produce a new map with no digon countries. If we can color the new map with 5 colors, we can color the old one with 5 colors also, because the tiny circular countries do not cover completely any of the boundary segments in the old map. In other words, color the new map, then let the tiny countries shrink back to the vertices they cover, and you will have the original map colored with 5 colors, with all countries contrasting where they border on each other. *Therefore, to prove the 5-color theorem, it will be sufficient to consider maps consisting entirely of polygon countries.*

FIGURE 5. If a map contains digon countries, it may be converted to a map without such countries by adding a tiny circular country in place of one vertex of each digon. If the new map can be colored with 5 colors, so can the original one. (Color the new map, then imagine that each circular country shrinks back to the vertex it covered.)

To summarize, we have reached the following four verdicts:

(1) We will temporarily assume that no country in our map contains a hole. In consequence, we may assume that every country (viewed as a face of a graph) is a polygon or a digon.

(2) Every vertex of the map has at least 3 edges (boundary segments) emanating from it.

(3) We may assume that the map viewed as a graph is connected.

(4) We may assume (over and above Verdict 1) that the map contains only polygon countries.

Now, if we turn back to the Descartes-Euler formula, Theorem 3 in the preceding chapter, we see that conditions (1), (3), and (4) above guarantee that the formula is applicable to the kind of map we are studying. We can now make use of the Descartes-Euler formula

$$F - E + V = 1$$

when the time is ripe.

EXERCISES

1. In the maps below straighten the boundary segments as much as you can, so that the maps become easier to think of as graphs.

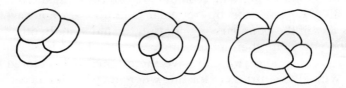

2. (Just to keep in practice.) Color the maps in Exercise 1 with as few colors as possible.

3. Draw a plane map of seven or more countries satisfying conditions 1 through 4. Count the numbers F, E, and V of countries, boundary segments, and vertices in the map, and check that

$$F - E + V = 1.$$

4. Draw a picture of a monogon country together with a country which borders on it.

5. Draw an example of a graph two of whose vertices have exactly two edges emanating from them. Smooth out the boundary segments at the two 2-edge vertices, so that the graph becomes a legitimate map. (If you wish, smooth out all the boundary segments so that the graph becomes easier to think of as a map.)

6. Draw an example of a map which is not connected when viewed as a graph.

7. Draw a map with ten countries, of which two are digons. Draw a copy of the map with one vertex of each digon country replaced by a tiny circular country. Color the new map with no more than 5 colors. Then color the original map by letting the two tiny circular countries shrink back to the vertices they cover.

3. Application of the Descartes-Euler Formula

For brevity, let us agree to call any map satisfying conditions 1 through 4 of the preceding section an *ordinary map*. We will use the Descartes-Euler formula to prove that every ordinary map has at least one country with relatively few edges or boundary segments. See Figure 6.

3. APPLICATION OF THE DESCARTES-EULER FORMULA

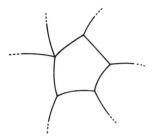

FIGURE 6. Any ordinary map contains a country which is bounded by at most 5 edges.

THEOREM 1. Any ordinary map contains a country that is bounded by at most 5 edges.

Proof. By condition 2, every vertex of the map has 3 or more edges emanating from it. Now, to every edge there are two ends. If there are E edges all told, then there are a total of $2E$ ends, clustered at least 3 at a time at the vertices. Therefore, the number V of vertices is at most equal to

$$\frac{2E}{3}.$$

Since

$$F - E + V = 1,$$

we have

$$F - E + \frac{2E}{3} \text{ is greater than or equal to } 1.$$

Equivalently,

$$F \text{ is greater than or equal to } 1 + \frac{E}{3}. \tag{1}$$

Now, let n denote the smallest number of edges possessed by any country in the map. Since each edge is shared by at most two countries, the total number E of edges must be at least equal to

$$\frac{nF}{2}.$$

Equivalently,

$$\frac{2E}{n} \text{ is greater than or equal to } F.$$

Combining this last statement with (1), we have

$$\frac{2E}{n} \text{ is greater than or equal to } 1 + \frac{E}{3},$$

$$2E \text{ is greater than or equal to } n\left(1 + \frac{E}{3}\right),$$

$$n \text{ is less than or equal to } \frac{2E}{1 + \frac{E}{3}}.$$

Now

$$\frac{2E}{1 + \frac{E}{3}} = \frac{6E}{E + 3}$$

$$= 6 - \frac{18}{E + 3}.$$

No matter what the number E of edges is, the number

$$6 - \frac{18}{E + 3}$$

is less than 6. Therefore n, the least number of edges possessed by any country, is at most 5. This completes the proof.

The reason we want to know there is a country with relatively few edges bounding it is so that we can do some coloring manipulations at that particular country. We will be ready to do the coloring as soon as we show that we can focus our attention on maps simpler yet than plain, old "ordinary" maps.

EXERCISES

1. Draw an ordinary plane map with nine countries, and count the numbers E and V for your map. Use your map as an example to illustrate the proof of Theorem 1.

2. Would you believe there is an ordinary plane map each of whose countries is bounded by 5 edges? Draw such a map. [*Hints:* (1) No map with fewer than 11 countries has this property. (2) If such a map has exactly 11 countries, each country must be bounded by exactly 5 edges. (3) Seek inspiration from the regular dodecahedron.]

4. Reduction to the Extra Ordinary Case

Let us agree to call a map an *extra ordinary map* if in addition to being ordinary it satisfies the condition

(5) Every vertex of the map has exactly 3 edges emanating from it.

For instance, the only vertex in the entire map of the United States that does not satisfy condition (5) is the "four corners" vertex at the meeting of Arizona, Colorado, New Mexico, and Utah.

THEOREM 2. If every extra ordinary map can be colored with 5 colors, then all ordinary maps can be colored with 5 colors.

Proof. Consider any ordinary map. Replace each vertex from which more than 3 edges emanate with a tiny circular country, small enough that it does not cover or touch any other vertex. See Figure 7.

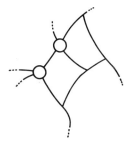

FIGURE 7. An ordinary map containing vertices from which more than three lines emanate can be converted to an "extra ordinary" map by replacing each such vertex with a small circular country.

The result of this replacement is a map which is extra ordinary, because all vertices with more than 3 edges emanating from them have been blotted out, and the tiny circular countries that cover them have only 3-edge vertices on their boundaries. Furthermore, if two countries border on each other in the original map, they still border on each other in the new map, because the circular countries added do not completely cover any edge.

Assuming then that every extra ordinary map can be colored with 5 colors, color the new map. Then the old map can be colored with 5 colors also, simply by using the same colors for those countries which appear in both maps. More picturesquely, color the new map with 5 colors, and then let the little circular countries shrink back to the vertices they cover as in Figure 8. The result will be the original map, colored in contrasting colors, since the shrinking process did not uncover any edges not present in the new map.

Therefore, we now only need to prove that every extra ordinary map can be colored with 5 colors. This completes the proof of Theorem 2.

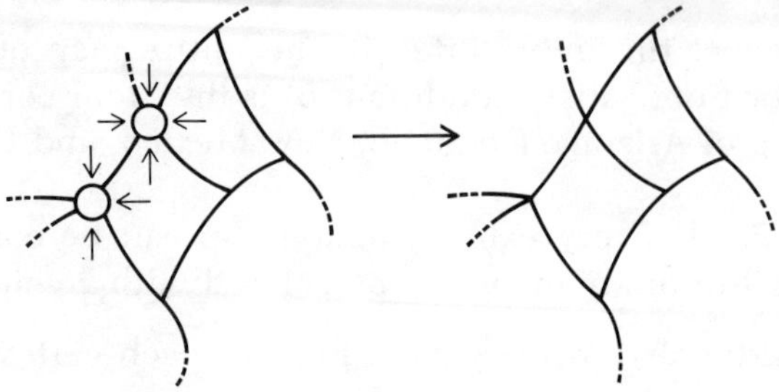

FIGURE 8. Once the extra ordinary map has been colored with 5 colors, one can obtain a 5-coloring of the original map simply by shrinking the tiny circular countries back to the vertices they were covering.

EXERCISE

1. Draw a 10-country ordinary plane map which is not extra ordinary. Convert it to an extra ordinary map by the process described in the proof of Theorem 2. Color the extra ordinary map with at most 5 colors. Then color the original map by letting the tiny circular countries shrink back to the vertices they cover.

5. Proof of the 5-Color Theorem for Ordinary Maps

THEOREM 3. Every ordinary map can be colored with 5 or fewer colors.

Proof. By Theorem 2, all we need to do is carry through the proof for an extra ordinary map. Let us take any extra ordinary map at all and locate within it a country A which is bounded by the least number of edges of any country in the map. By Theorem 1, that country is deformable to a triangle, a 4-sided figure, or a 5-sided figure. Since each vertex of our map has exactly 3 edges emanating from it, the country A and the edges emanating from its boundary must be deformable to one of (a), (b), or (c) in Figure 9.

5. PROOF OF THE 5-COLOR THEOREM FOR ORDINARY MAPS

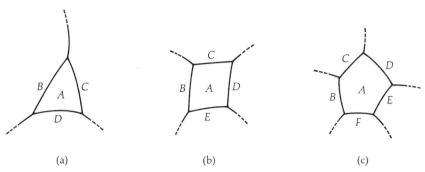

(a)   (b)   (c)

FIGURE 9. The country A with the least number of edges, together with the edges emanating from its boundary, is deformable to one of these three figures: (a) Triangular. (b) 4-Sided. (c) 5-Sided.

Case (a). In case the country A can be deformed to the triangle of Figure 9(a), consider the new map in Figure 10 obtained by shrinking country A to a point. The new map has one country fewer than the original one, and it is still extra ordinary. If the new map can be colored with 5 colors, then so can the original one, because the country A can be given either of the 2 colors not given to countries B, C, and D. Therefore, in this case we have reduced the coloring problem to the same problem for a map with one country fewer. We will now show that the same reduction to a map with fewer countries can be accomplished in the other two cases.

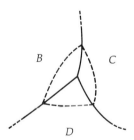

FIGURE 10. In case the country with the smallest number of edges is triangular, consider a new map obtained by shrinking that country to a point. If the new map can be colored with 5 colors, then so can the original one, because the region A can be given either color different from the colors of B, C, and D.

Case (b). In case the country A can be deformed to the 4-sided Figure 9(b), we investigate the pairs of opposite countries B, D and C, E. Either B and D have no point in common or else they loop around to touch each other or to form a single country. But in case B and D

do loop around as in Figure 11, the loop they form divides country C from country E. Therefore, at least one of the pairs B, D and C, E is a pair of different countries with no border in common. Suppose, changing the labels if necessary, that C and E are different countries.

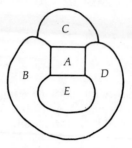

FIGURE 11. In case the country A with the smallest number of edges has four edges, two of the countries bordering on A may loop around to border on each other or even join as a single country. If this happens, however, the other two countries, C and E in this figure, will *not* border on each other.

We form a new map by joining countries C and A and removing the vertices labeled 1 and 2 in Figure 12. The new map has one country fewer than the original one, and it is still extra ordinary. Suppose the new map can be colored with 5 colors. Then the old map can be colored with 5 colors by replacing the boundary between A and C and then changing the color of the region A so that it is different from the colors which were given to B, A–C, D, and E. Since there are 5 colors available, this color change can be made.

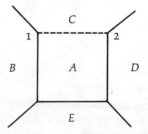

FIGURE 12. In case the country A with the smallest number of edges has four edges, consider the map obtained by removing the boundary between the countries A and C. If the new map can be colored with five colors, then so can the original one by recoloring the country A with a color different from those of B, A-C, D, and E.

Case (c). In case the country A can be deformed to the 5-sided Figure 9(c), we can use the same argument we gave in Case (b) to

5. PROOF OF THE 5-COLOR THEOREM FOR ORDINARY MAPS

conclude that among the 5 countries B, C, D, E, and F there must be a pair that are different countries and that do not touch each other. Changing the labels if necessary, let us call the pair C and E.

We form a new map by joining countries C, A, and E and removing the 4 vertices labeled 1, 2, 3, and 4 in Figure 13. The new map has two countries fewer than the original one, and it is still extra ordinary. Suppose the new map can be colored with 5 colors. Then the old map can be colored with 5 colors by replacing the boundaries between A and C and between E and C and then changing the color of the region A so that it is different from the colors which were given to B, C–A–E, D, and F. Since there are 5 colors available, this color change can be made.

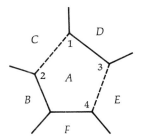

FIGURE 13. In case the country A with the smallest number of edges has five edges, consider the map obtained by removing the boundaries between C, A, and E. If the new map can be colored with five colors, then so can the original one by recoloring the country A with a color different from those of B, C-A-E, D, and F.

In each of the three cases above, for verbal simplicity we spoke of the regions B, C, D, E, and F as if they could have been nothing other than countries in the map and therefore to be colored. Actually, one and sometimes two of the regions may be the region outside of the map. If that is true, however, the arguments are still valid. In fact, in picking the new color for A there is then more freedom, since it need not contrast with so many countries.

In each case of the number of edges of country A, we have shown that the map-coloring problem can be reduced to the same problem for an extra ordinary map containing fewer countries. The problem for that map can then be reduced in turn and so on until the original problem is finally boiled down to the problem of coloring an extra ordinary map having 5 or fewer countries. Since the latter problem can be solved by giving each country a different color, it follows that any map can be colored with 5 or fewer colors. This completes the proof of the theorem.

106   CHAP. 5   MAP COLORING

EXERCISES

1. Explain why in Case (b) of the proof of Theorem 3, we must be careful to pick the countries C and E to be different countries. [*Hint:* Refer back to the conditions which make up "ordinariness."]

2. Explain why in Case (c) of the proof of Theorem 3, we must be careful to pick the countries C and E not only to be different countries but also to be not even touching each other.

3. Draw an extra ordinary map with 7 countries, and use the proof of Theorem 3 to reduce the problem of coloring it to the problem of coloring a map with 6 countries. Draw the 6-country map, and then use the proof of Theorem 5 to reduce the problem of coloring it to the problem of coloring a map with 5 countries. Draw the 5-country map, color it, and then work up to a coloring of the 7-country map.

6. Maps with Ring-Shaped Countries, Maps on the Surface of a Sphere, Maps on the Surface of a Doughnut

Now that we have established the 5-color theorem for ordinary maps in a plane, we would like to show that the same result holds if the map contains ring shaped or Swiss cheese shaped countries. We would also like to show that the same 5-color theorem holds for maps on the surface of a sphere (like a map of the earth we live on). Each of these additional results is very easy to prove. All we have to do is make a few appropriate slits and holes in the new kinds of maps, and they will be reduced to the kinds of maps for which the 5-color theorem is known to hold.

THEOREM 4.   Any map drawn in a plane can be colored with 5 or fewer colors.

Proof.   In each ring shaped or Swiss cheese shaped country, cut as many narrow openings as are necessary to open the holes up, but do not cut so many as to divide the country into more than one piece. See Figure 14. Make each opening sufficiently narrow that each of its ends butts up against only a middle portion of an edge of a country. Treating each opening as a country, color the new map. By Theorem 3, the coloring can be done with 5 or fewer colors. Now change the color of each opening to the color of the ring country in which the opening was made. Or, more picturesquely, think of the narrow incisions as healing. The result is the required coloring of the original map. This completes the proof.

FIGURE 14. In this map we have cut a narrow opening in the ring shaped country to open the hole up. We know that the map with the opened up ring can be colored with 5 colors. Therefore the original map can be colored with 5 colors also, simply by changing the color of the narrow slit to the color of the rest of the ring.

THEOREM 5. Any map drawn on the surface of a sphere can be colored with 5 or fewer colors.

Proof. Make a tiny hole in the spherical surface within any one country. This hole changes no bordering relationships, but it allows us to deform the spherical map until it is flat—a plane map. See Figure 15. By Theorem 4, the plane map can be colored with 5 or fewer colors. Therefore, since the boundary relationships are the same, the spherical map can be colored in the same way with 5 or fewer colors. This completes the proof.

The remaining portion of this section's title is somewhat curious. A map on a doughnut? Well, it does not do any harm to think of such a thing. It may even help us better understand the proofs we have just completed, by showing us where they must be changed when we encounter a doughnut instead of a plane or sphere. The places where the proofs must be changed are those steps which make crucial use of the topological properties of the plane.

To begin, let us say that 5 colors are not sufficient to color maps on the surface of a doughnut. What is there about the doughnut that prevents the proofs we have just gone through from working? Well, first of all, the Descartes-Euler formula, used in proving Theorem 1, does not necessarily hold for graphs not drawn in a plane. As a matter of fact, there is a similar formula for graphs with polygon faces drawn on the surface of a doughnut, but it reads

$$V - E + F = 0.$$

We will not prove this formula, but you might like to draw a graph on a doughnut and see for yourself that it holds for your particular graph. (You might also like to refer to Exercises 8 and 10 of Chapter 4, Section 5.)

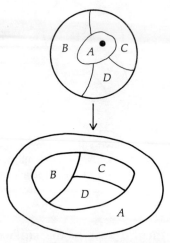

FIGURE 15. Transformation of a 4-country map on the surface of a sphere to a 4-country map in a plane. A tiny hole is cut in country A, and then the spherical surface is stretched until it lies flat. (The edge of the hole stretches to form the outer boundary of the ring country A.)

Therefore, we cannot depend upon Theorem 1 holding for a map on the surface of a doughnut. What is true is that some country in any map on the surface of a doughnut has 6 or fewer edges.

A second place where we used a special property of the plane was in Cases (b) and (c) of the proof of Theorem 3. On the surface of a doughnut, it is possible to have the situation seen in Figure 16, where both B and D are a single country *and* C and E are a single country. Therefore that part of the argument falls apart.

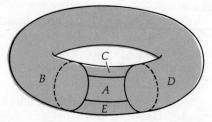

FIGURE 16. On the surface of a doughnut, it is possible to have a four-sided country which borders on only two countries.

It turns out that 7 is the magic number for coloring maps on the surface of a doughnut. P. J. Heawood not only proved that any map on

6. MAPS WITH RING-SHAPED COUNTRIES

the surface of a doughnut can be colored with 7 or fewer colors, but he also gave an example of a 7-country map in which each country borders on the other 6. Therefore, some maps on the surface of a doughnut actually require 7 colors. An example of Heawood's 7-country–7-color map is shown in Figure 17. When the ends of the rectangle are pasted together to produce a doughnut shaped surface, every country will border on every other country.

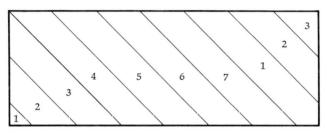

FIGURE 17. Imagine this rectangle to be made of some extremely elastic material. If it is first rolled over lengthwise to form a tube and then turned so that the ends of the tube are joined, the resulting doughnut-shaped surface will consist of exactly seven colored regions, each one bordering on the other six.

Remember we do not have such complete information for plane maps. We know that no plane map will ever require more than 5 colors, but no one has *ever* found a plane map requiring 5 colors. (It has been shown that 4 colors are sufficient for any map in the plane having 39 or fewer countries. Therefore, if some map does require 5 colors, it must contain 40 or more countries and presumably will be difficult to construct.)

Therefore Francis Guthrie's original question, "Are 4 colors sufficient?" has not been fully answered. Most mathematicians seem to be betting that 4 colors are sufficient, so the problem goes by the name of the *Four-Color Problem*.

Gerhard Ringel and J. W. T. Youngs have recently solved the map-coloring problem for all surfaces other than the plane and sphere (surfaces, for example, like a doughnut with as many holes as you please), so the problem for maps on the simplest kind of surface is the only one not yet answered.

EXERCISES

1. Draw an example of a plane map with at least 3 countries containing holes. Put at least 4 holes in one of the countries. Draw the incisions called for in the proof of Theorem 4.

110   CHAP. 5 MAP COLORING

2. Draw a 6-country map covering the surface of a sphere, and then draw a plane map to which it would deform in the proof of Theorem 5.

3. Why is it necessary to prove Theorem 4 before proving Theorem 5?

4. If a small hole is cut in one face of a cube, the remainder of the cube can be stretched until it lies flat in the plane. Draw a map in the plane whose countries are the 6 faces of the cube after the stretching.

5. Repeat Exercise 4 for the regular tetrahedron, octahedron, dodecahedron, and icosahedron.

6. Calculate $F - E + V$ for the graph on the doughnut in Figure 16. (Be sure to count all three faces.)

7. Draw a map of 9 countries covering the surface of a doughnut and color it with 7 or fewer colors. [*Suggestion:* You need not follow this exercise literally; a ring of paper works just as well and is not so perishable.]

8. Calculate $F - E + V$ for your map in Exercise 7.

9. Make a sketch of 5 solid figures in space, each of which touches the other 4 along a portion of a surface.

10. Repeat Exercise 9 for 10 solid figures, each touching the other 9 along a portion of a surface. Is there any limit to the number of solid figures that can be constructed in space, each one touching all the others along a portion of a surface?

11. Assume that any graph with polygon faces drawn on the surface of a doughnut satisfies

$$V - E + F = 0.$$

Prove that if every vertex of the graph has at least three edges emanating from it, then some face of the graph is bounded by 6 or fewer edges.

12. Give an example of a graph with polygon faces drawn on the surface of a doughnut none of whose faces is bounded by fewer than 6 edges. [*Hint:* See Figure 17.]

CHAPTER SIX

Miscellaneous Topics

1. Introduction

This chapter and the next one are collections of mathematical recreations or curiosa. There are two reasons for including such a miscellany in this book. First, much of mathematics has arisen simply because men applied their wits to puzzles or games or curiosities. Second, a bright little collection of tricks will form a pleasant interlude between the more serious and thorough-going chapters which preceded and which will follow.

2. Fiftieth Anniversaries

If you have ever kept track of the days of the week on which Christmas (or any other anniversary) appears from one year to the next, you know there is a regularity to its appearance. With the passing of each ordinary year, Christmas occurs 1 weekday later than it did the previous year. With the passing of a leap year, Christmas occurs 2 weekdays later than it did the previous year.

The reason is that each year has just a bit more in it than 52 weeks. An ordinary year is 52 weeks plus 1 day; a leap year is 52 weeks plus 2 days. Thus, in 365 days from December 25, the week calendar will have rolled by 52 times and then advanced 1 more day. In 366 days, the motion of the week calendar will have been 52 complete cycles

112   CHAP. 6 MISCELLANEOUS TOPICS

plus 2 more days. A remarkable gentleman I met while I was roaming around the Grand Canyon used these observations to explain a phenomenon he had noticed: That fiftieth anniversaries will commonly fall on the same day of the week as the event itself or one day behind (with certain exceptions).

I am giving his explanation not because it is profound but rather for the opposite reason. It is a beautiful elementary piece of reasoning closely related to one of the key ideas in number theory—that of studying remainders of numbers after they have all been divided by a single natural number. (Here the divisor is the number 7.) This gentleman devised his explanation with no formal study in mathematics beyond what he had had in high school in the 1910s.

Suppose for simplicity that an event takes place any day of the year except February 29th. If there were no such thing as a leap year, then the weekday on which the anniversary of the event falls would advance by 1 day as each year passed. Over a 50-year period, this would be an advance of 50 days, which is the same as an advance of 1 day, since 50 days equals 7 weeks plus 1 day. But the added February 29ths change this; each time a February 29th is stuck in, the day of the week on which the anniversary falls advances by two instead of by one. Therefore, in 50 years the correct amount of advance is one plus the number of February 29ths that occur between the event and its fiftieth anniversary. If February 29ths occur every 4 years (exception to this will be noted shortly), there will be either 12 or 13 February 29ths in between.

In case there are 13, then the advance in weekdays of the fiftieth anniversary is $1 + 13 = 14$, or 2 weeks exactly, which is the same as no advance. In this case the fiftieth anniversary will fall on the same day of the week as the event itself. In case there are 12 February 29ths, the advance in weekdays is $1 + 12 = 13$, or 1 week and 6 days, or 2 weeks less 1 day. In this case, depending on how you wish to think of it, the fiftieth anniversary of the event will fall 1 weekday behind or 6 weekdays ahead of the day of the week on which the event occurred.

The complete rule for determining which years are leap years is complicated:

(1) Only years whose numbers are evenly divisible by 4 are eligible to be leap years, but not all of them are leap years. Such a year *is* a leap year unless its number is evenly divisible by 100.

(2) Years with numbers divisible by 100 but not by 400 are *not* leap years.

(3) Years with numbers divisible by 400 *are* leap years.

For example, 1600 and 2000 were and will be leap years, but 1700, 1800, and 1900 were ordinary years.

2. FIFTIETH ANNIVERSARIES

Let us see how the leap years fall in a specific example. On Monday, May 10, 1869, the golden spike joining the Union Pacific and the Central Pacific railroads was driven at Promontory, Utah. At that moment the United States acquired its first transcontinental railroad. Between that event and its fiftieth anniversary, there were only eleven February 29ths (in the years 1872, 1876, 1880, 1884, 1888, 1892, 1896, 1904, 1908, 1912, and 1916). We thus have the exceptional case of a leap year being skipped in 1900, so the fiftieth anniversary occurred $1 + 11$ weekdays advanced from the Monday on which the event occurred. That is, the fiftieth anniversary occurred on Saturday, May 10, 1919, Saturday being the weekday 12 days advanced from Monday (or, equivalently, 2 days behind Monday).

The one hundredth anniversary of the event occurred on Saturday, May 10, 1969, the same day of the week on which the fiftieth anniversary occurred, because there were thirteen February 29ths between the fiftieth anniversary and the one hundredth anniversary.

Now that we know the exception to the rule that leap years come every 4 years, we can say for certain that between any event and its fiftieth anniversary there will be exactly 11, exactly 12, or exactly 13 February 29ths. Thus, the day of the week on which a fiftieth anniversary falls will be respectively 2, 1, or 0 days behind the day of the week on which the event itself took place.

The omission of the three out of four century leap years from the calendar was decreed in 1582 by Pope Gregory XIII, to help correct the tendency of the Julian calendar (introduced by Julius Caesar) to run behind the solar year. In its sixteen centuries of use, with a leap year every 4 years, the Julian calendar had fallen 10 days behind the solar calendar. To close this gap, the 10 days from March 11 to March 20 were omitted from the year 1582.

The various countries adopted the changes with a speed inversely related to their distance (geographical, political, and religious) from the Vatican. England and her colonies did not make the changes until 1752, when they dropped 11 days from the year (because the year 1700 had passed). When George Washington was born, the calendar on the wall read February 11, but with the change he celebrated his birthday on the 22nd, and now so do we. Russia made the change only after the Communists took power; the "October revolution" which brought them to power actually occurred in November according to the Gregorian calendar.

EXERCISES

1. How many days had to be dropped from the calendar when Russia converted to the Gregorian calendar?

114   CHAP. 6   MISCELLANEOUS TOPICS

2. What is the rule for determining the day of the week of a one hundredth anniversary?

3. What is the rule for determining the day of the week of a four hundredth anniversary?

4. As of this writing, some Eastern Orthodox congregations still determine their feast days by the Julian calendar. On what date in the Gregorian calendar do they celebrate Christmas?

5. On what day of the week were you born? [*Hint:* Work backward from your birthday this year. If you were born on February 29th, do the calculations for February 28th and then advance the day of the week by one at the end.]

6. Pick out 3 fiftieth anniversaries that have occurred between 1940 and now. Then check back in the library collection of newspapers, magazines, or calendars to verify the rule for determining the day of the week of the anniversary. In each case list all February 29ths occurring between the event and its fiftieth anniversary.

7. National elections for the presidency of the United States are always held on the first Tuesday of November, in years whose numbers are divisible by 4. In 1968 the presidential election took place on November 5th. Calculate the dates of the presidential elections in the years 1964, 1972, 1976, 1980, and 1984. Can you give a rule for determining the date of each presidential election in this century?

8. Find out the latest estimate of the length of a solar year from an encyclopedia or other source. How many years does it take for the Gregorian calendar to become one day in error from the solar calendar? In which direction is the error?

9. Prove that in the Gregorian calendar the thirteenth day of the month falls more often on Friday than on any other single day of the week. [*Hint:* You should make use of what you learned in Exercise 3.]

3. Knots, Braids, and a Surprise

There is one little area of topology into which a person can go far on intuition alone, but which requires a lot of theory to be studied rigorously. That is the general area having to do with knot-tying and braiding.

3. KNOTS, BRAIDS, AND A SURPRISE

Suppose you have tied two knots in two pieces of rope, an overhand or "simple" knot in one and a slip knot in the other. The two knots are shown in Figures 1 and 2. A topologist thinks of these two knots as being different for the following reason: If the two ends of the rope containing the overhand knot are spliced together (along the dotted lines), it is impossible to untie that knot, no matter how long the rope, no matter what contortions you put it through. On the other hand, if the two ends of the rope containing the slip knot are spliced together, the slip knot can be "slipped" until it dissolves, leaving a plain loop of rope.

FIGURE 1. An overhand knot.

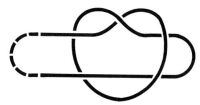

FIGURE 2. A slip knot.

Since the one knot can be maneuvered into a plain loop while the other cannot, it follows that it is impossible to maneuver either of the two knots of Figures 1 and 2 into the other, so long as the ends of the ropes are spliced. For that reason, the two knots are said to be *inequivalent*.

The equivalence relation defined among knotted loops by the "maneuverability property" splits up the various possible knots into classes of equivalent knots. You might remember from earlier days that there are many knots that can be tied in the middle of a rope, without the ends being available. Any knot that can be so tied can also be untied without the ends being used, so any such knot tied in a large enough loop can be dissolved into a plain, unknotted loop. Thus, there are many knots equivalent to a simple loop.

How can one tell if one complicated knot is equivalent to another? That is a very difficult problem. If you have ever watched someone unravel a piece of unfinished knitting, you know that ordinary knitting without the finishing locking knot can be completely dissolved merely by pulling on one end. Yet if even once the ball of yarn is passed through a loop, creating an almost imperceptible alteration in the whole fabric, then the dissolution is no longer possible.

Braiding is in one way a generalization and in another way a specialization of knot tying. Braiding can be done with as many strands as a person wishes to use—as opposed to the single rope used in knot tying. However, according to the definition of a braid, the person doing the braiding must only pass the various strands over and under the free ends of the other strands. That is, one is not allowed to take a strand and weave it in and out of the portion of the braid already created; braiding must proceed in one direction only, away from tied ends of the strands.

For any number n of strands, there is an especially simple, regular kind of braid, called the *standard braid*. It is begun by taking the strand on the extreme right and passing that strand over, then under, then over, then under, ... all the strands to its left until it becomes the left-most strand. The braid can then be continued indefinitely by repeating this step with each new strand that appears on the extreme right. Figure 3 shows the first two steps performed on a 5-strand braid. In the case $n = 3$, the standard braid is the same as the ordinary pigtail braid, although most people use a slightly different technique to produce the pigtail braid.

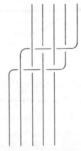

FIGURE 3. The first two steps in making a standard braid on five strands.

In the case $n = 2$, you can see that the standard braid consists merely of two strands twisting around each other in a spiral. That is, in the special case $n = 2$, the standard braid can be produced even if the free ends of the strands have been tied together. That, of course, is not very surprising. What is surprising to almost anyone seeing it for the first time is *that very same thing being done* in the case $n = 3$. Yes, the standard braid on 3 strands can be produced even if the free ends of the 3 strands have been tied together. You might like to try it yourself, following the directions given below, using a thin strip of paper slit into three strands but with the ends left joined.

3. KNOTS, BRAIDS, AND A SURPRISE   117

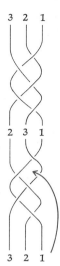

FIGURE 4. Pass bottom end through indicated spot from front side, and tangle will reduce to that in Figure 5.

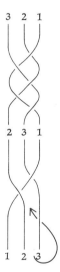

FIGURE 5. Pass bottom end through indicated spot from back side, and tangle will reduce to strand 3 lying under strand 2, which completes the braid cycle.

The best way of doing the braiding is to go through almost one full cycle as shown in Figure 4, and then push the joined bottom ends through the tangle two times (at just the right places, shown in Figures 4 and 5). The tangle will vanish, leaving only an overlap of one strand on another, which completes the cycle. You can repeat the process as many times as you have room for. When you run out of maneuvering room, there will be a small portion left unbraided. This can be filled in by the expedient of loosening the braiding above and then pulling it down until the braid is uniform from one end to the other.

This technique, sometimes called split-strap braiding, was discovered long before the word "topology" was heard of. People just discovered it, undoubtedly by trial and error. It is not too surprising that such a trick should have been found — given the intelligence and curiosity of man, plus the number of centuries he has been experimenting with the world about him. Look how many complex folk medicines and treatments have been found to have genuinely curative properties, often after being regarded as useless by the first "civilized" people to see them.

EXERCISES

1. Draw the knot that is the mirror image of Figure 1. Tie this knot in a piece of string, splice the ends (or join them somehow), and then see if you can maneuver it into the knot of Figure 1.

2. In a scout manual, mountain climber's manual, or whatever, find three different named knots that can be tied in the middle of a rope without using the ends.

3. Split-strap braid a strip of paper, just to get the feel of it. Notice that the individual strands are not twisted about themselves in the end product.

4. Split-strap braid a strip of paper with 2 strands instead of 3. Can you do it without having at least one strand twist about itself?

5. Explain as best you can the difference between 2- and 3-strand braiding, that it should be possible to do Exercise 3 without twisting the strands but not Exercise 4.

4. Matching Birthdays

In a group of 30 people, what would you say is the probability that at least one pair of them will have the same birthday? Thinking of the number 30 as the number of days in a month, which is approximately 1/12 of a year, a person might first guess that the probability is about

1/12. (That is, of many different groups of 30 people each, one might expect that about 1/12 of the groups would contain pairs of people with the same birthday.) In reality, the probability that out of a group of 30 at least one pair will have the same birthday is slightly over 7/10, and in a group of only 23 people the probability is already over 1/2. Anything so contrary to intuition really deserves to be investigated.

The problem of computing the chance of having a pair of people with the same birthday is typical of many probability problems in that it is extremely difficult to compute directly the probability of the event in question but relatively easy to compute the probability of the opposite event. That is, we will find it much easier to answer first the opposite question: What is the probability that no two people in a group have the same birthday?

Suppose we have a group of n people, numbered from 1 to n. The condition that no two people have the same birthday is expressible as follows:

(2) Number 2's birthday is different from Number 1's birthday, *and*

(3) Number 3's birthday is different from the two birthdays of Number 1 and 2, *and*

(4) Number 4's birthday is different from the three birthdays of Numbers 1, 2, and 3, *and* . . .
and

(n) Number n's birthday is different from the $n-1$ birthdays of Numbers 1 to $n-1$.

Now for Statement (2) to be true, Number 2's birthday can be on any day of the year except one. Therefore, the probability of Statement (2) being true is 364/365. For Statement (3) to be true, Number 3's birthday can be on any day of the year except two. Therefore the probability of Statement (3) being true is 363/365, the probability of Statement (4) being true is 362/365, the probability of Statement (5) being true is 361/365, and so on down to the probability of Statement n being true:

$$\frac{365-(n-1)}{365}.$$

The probability of *all* these statements being true — the probability that *no* two out of n people have the same birthday — is the product of all these probabilities:

$$\frac{364}{365} \times \frac{363}{365} \times \frac{362}{365} \times \frac{361}{365} \times \cdots \times \frac{365-(n-1)}{365}. \tag{1}$$

Now we can begin to see something. If n is even moderately large, the fractions toward the right-hand side of this product will be moderately smaller than 1. Although each individual fraction may not be greatly different from 1, multiplying them all together will have a profound effect on the value of the product. Take, for example, $n = 23$. Then we have for the probability that no two people of the 23 have the same birthday:

$$\frac{364}{365} \times \frac{363}{365} \times \frac{362}{365} \times \cdots \times \frac{346}{365} \times \frac{345}{365} \times \frac{344}{365} \times \frac{343}{365}. \qquad (2)$$

You can work out for yourself that the product of the last four fractions alone is approximately .794. And then there are eighteen more fractions to be multiplied, although these become closer to 1 the farther to the left they appear.

It should take you about 3 minutes to calculate the entire product (2) on a slide rule. Although the accuracy of the slide rule answer will not be so wonderful, the result will still be convincing. With a desk calculator, with a computer, or with a pencil and a lot of patience, one can obtain the answer to greater accuracy: In a group of 23 people, the probability that no two will have the same birthday is approximately .493. Thus, the probability that *some* two people in a group of 23 will have the same birthday is

$$1 - .493 = .507.$$

Figure 6 is a graph of the probabilities that in a group of size n there will be at least two people with the same birthday. The graph only covers groups of sizes 1 up to 70, because for groups of more than 70 people the probability is so close to 1 that the difference is indistinguishable on the size of graph printed here. Notice that even though the graph necessarily rises in discrete jumps (since there are no groups of $23\frac{1}{2}$ people, and so on), still there is a great regularity to the overall rise.

We will finish this section with a bit of philosophy. You undoubtedly have noticed that we have not taken into consideration birthdays occurring on February 29th. We have ignored them for the sake of simplicity; including them makes the computations *much* more involved, although when the smoke clears one finds that the probabilities have decreased only slightly from those of Figure 6.

Another phenomenon has the opposite effect. All the factors in formula (1) are based on the assumption that all days of the year have about the same number of people born on them. This is really not the case, simply because conceptions occur more often on some days than on others. (Speculation on the part of the reader is invited.) Table

1 shows the number of births in the United States in each month of 1966 and the average number of births per day in each of these months.

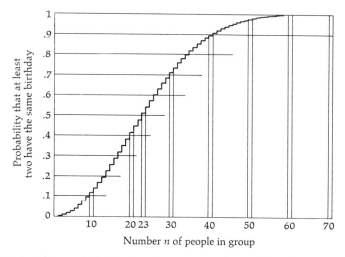

FIGURE 6. The probability that, in a group of n people, at least two people will have the same birthday.

TABLE 1. For the year 1966 in the United States: The number of births in each month and the average number of births per day in each month.

Month	Births	Average Number of Births per Day
January	293,850	9479
February	273,904	9782
March	303,420	9788
April	286,914	9564
May	292,824	9446
June	292,526	9751
July	310,550	10,018
August	321,304	10,365
September	319,234	10,641
October	312,942	10,095
November	296,458	9882
December	302,350	9753

This nonuniformity of births around the year makes it "easier" for two people in a group to have the same birthday. You can see how the probabilities of having matching birthdays would dramatically increase if all births should occur in a single month. The slight tendency in this direction (of some months having more than their share of births) also produces increases in the probabilities, but less spectacular ones. To the best of my knowledge, no one has ever done an empirical study to see just how large the increases are.

EXERCISES

1. How many people must there be in a group in order to have probability greater than $\frac{1}{2}$ that at least two were born on the same day of the week?

2. How many people must there be in a group in order to have probability greater than $\frac{1}{2}$ that at least two were born in the same month? (Assume for simplicity that the same number of births occur in each month.)

3. Draw graphs similar to Figure 6 for Exercises 1 and 2.

4. How many presidents of the United States have there been? From Figure 6, what is the probability that in a group so large at least two people have the same birthday?

5. How many presidents of the United States are presently alive? From Figure 6, what is the probability that in a group as large as the deceased presidents of the United States at least two people will die (or died) on the same day of the year?

6. Two presidents of the United States were born on November 2. Who were they? Three presidents of the United States died on July 4. Who were they?

7. Sample the birthdays of people you pick at random (in your dormitory, from *Who's Who*, or from some other source you happen to think of) until your sample includes two people with the same birthday. How many people did you have to pick?

8. What was the probability in doing Exercise 7 that you would have wound up with 23 or fewer people, assuming that birthdays are evenly distributed around the year? Describe how you might develop a class research project to see how much effect the varying birth rates in the different months have on the probability of finding within a given group a pair of people with the same birthday.

5. Casting Out Nines

After performing a really long addition or multiplication with pencil and paper, have you ever wished you could check your results without going through all the work again? There is a way, based on a couple of elementary number-theoretic results. It does not catch all possible mistakes, but it will detect the vast majority of them.

Pick a natural number n larger than 1. We will constantly be taking remainders of natural numbers after division by n, so let us adopt a special notation for remainders. We denote by $R_n(a)$ the remainder of a when it is divided by n.

For example, if $n = 7$,

$$R_7(16) = 2, \qquad R_7(45) = 3,$$

and

$$R_7(16 + 45) = R_7(61) = 5.$$

We notice that two of these remainders, 2 and 3, add up to the third one, 5. Is it just an accident that $R_7(16 + 45)$ is the sum of $R_7(16)$ and $R_7(45)$, or is this an example of a relationship which holds in general? Well, the sum of two remainders may be too large to be a remainder itself, so we cannot say simply that, "The remainder of a sum equals the sum of the remainders." However, the next best thing is always true:

THEOREM 1. Let a, a_i, b, and n be natural numbers. Then:

  (i) $R_n(a + b) = R_n(R_n(a) + R_n(b))$
  (ii) $R_n(a_1 + \cdots + a_k) = R_n(R_n(a_1) + \cdots + R_n(a_k))$
  (iii) $R_n(ab) = R_n(R_n(a)R_n(b))$.

For example, letting $n = 7$, $a = 13$, and $b = 32$, in formula (i), we have

$$R_7(13 + 32) = R_7(45)$$
$$= 3.$$
$$R_7(R_7(13) + R_7(32)) = R_7(6 + 4)$$
$$= R_7(10)$$
$$= 3.$$

You should make up similar examples to help you see the meaning of formulas (ii) and (iii).

Proof. We will write out the proof of formula (i) only. The other two proofs are very similar. By definition of remainders

$$a = q_1 n + R_n(a),$$
$$b = q_2 n + R_n(b),$$

and

$$a + b = q_3 n + R_n(a + b),$$

where q_1, q_2, and q_3 are natural numbers (or possibly 0). If we add the first two equations, we obtain an expression on the right-hand side which is equal to $a + b$. It therefore must be equal to the right-hand side of the third equation:

$$(q_1 + q_2) n + R_n(a) + R_n(b) = q_3 n + R_n(a + b).$$

This equation can be rewritten as

$$R_n(a) + R_n(b) = (q_3 - q_1 - q_2)n + R_n(a + b).$$

Now look very carefully at the right-hand side of this last equation, because it tells us exactly what we are trying to prove. The first term is a multiple of n, while the second term is 0 or a natural number less than n. That is, the second term is the *remainder* obtained when the left-hand side is divided by n:

$$R_n(a + b) = R_n(R_n(a) + R_n(b))$$

which is Equation (i). This completes as much of the proof as we will give.

The idea of checking arithmetic by means of remainders is this: If, after performing an addition of two natural numbers a and b, obtaining the answer c, we then find

$$R_n(c) \neq R_n(R_n(a) + R_n(b)),$$

then we *know* we have made a mistake. However, if we do get equality of remainders, we are not absolutely certain that our addition was correct. But we are sure that either our answer is the right one or the difference between our answer and the right one is some multiple of n. In case of an addition with more than two summands, the same checking technique works, since the more general Equation (ii) holds. Just add all the remainders of the various summands and compare the remainder of that sum with the remainder of your answer.

The same philosophy can be used to check multiplications, because of Equation (iii). Subtractions can be checked by checking the process of addition which reverses the subtraction. Division with remainder can be checked by checking the combined multiplication-addition which reverses the division.

However, all of the above theory is impractical *unless* there is some number n for which it is possible to find $R_n(a)$ quickly, without going through an actual division. Such a number does exist, namely 9. The short technique for finding $R_9(a)$ is called "casting out 9's." The underlying reason why there is such a technique is that

$$9 = 10 - 1$$

5. CASTING OUT NINES

and ten is the "base" of the decimal system. Had history given us some other base for our number system, we could still cast out something, the something being the number one less than the base—the largest single-digit number. The casting out 9's process is described in the following theorem:

THEOREM 2. The remainder of any natural number after division by 9 is the same as the remainder of the *sum of its digits* after division by 9.

Proof. Suppose the digits are $\cdots e, d, c, b, a$. That is, suppose the number in question is equal to

$$\cdots 10000e + 1000d + 100c + 10b + a.$$

Now, any power of 10 leaves a remainder of 1 when divided by 9, because any power of 10 can be written

$$10 \cdots 000 = 9 \cdots 999 + 1$$
$$= (1 \cdots 111)9 + 1.$$

Let x be any one of the single-digit numbers $0, 1, 2, \cdots, 8, 9$. Then by Equation (iii) of Theorem 1,

$$R_9(10 \cdots 000x) = R_9(R_9(10 \cdots 000)R_9(x))$$
$$= R_9(1 \cdot R_9(x))$$
$$= R_9(x).$$

Therefore by Equation (ii) of Theorem 1,

$$R_9(\cdots 10000e + 1000d + 100c + 10b + a)$$
$$= R_9(\cdots R_9(10000e) + R_9(1000d) + R_9(100c) + R_9(10b) + R_9(a))$$
$$= R_9(\cdots R_9(e) + R_9(d) + R_9(c) + R_9(b) + R_9(a))$$
$$= R_9(\cdots e + d + c + b + a).$$

The last step is possible because $R_9(x) = x$ for any digit x, except in the case $x = 9$. But in the case $x = 9$, $R_9(9) = 0$ is replaced by the digit 9, and this does not change the remainder of the sum

$$\cdots R_9(e) + R_9(d) + R_9(c) + R_9(b) + R_9(a)$$

when it is divided by 9. This completes the proof.

Unless the number we are dealing with is truly huge, summing the digits once or perhaps twice is sufficient to enable us to find the remainder of that number after division by 9. As an example, let us find $R_9(1{,}257{,}393)$.

$$R_9(1{,}257{,}393) = R_9(1 + 2 + 5 + 7 + 3 + 9 + 3)$$
$$= R_9(30)$$
$$= R_9(3)$$
$$= 3.$$

For the technique of checking by using remainders R_n to be practical, n must be a number such that mistakes differing from the true answer by a multiple of n are not common. Does $n = 9$ have this property? Well, mistakes in addition ordinarily consist of misadding a column, and the wrong column sum usually differs from the correct one by some small number like 1, 2, or 3. The remainder R_9 of a sum resulting from such a mistake will differ from the remainder of the true value by this same 1, 2, or 3, on account of Theorem 2. That is, a single mistake in addition almost never yields an incorrect answer differing from the true value by a multiple of 9.

The same is true of most errors in multiplication, with one notable exception: An error in *position* cannot be detected by casting out 9's, because it *never* alters the value of R_9. For example, try checking the following erroneous multiplication.

$$\begin{array}{r} 351 \\ \underline{405} \\ 1755 \\ \underline{1404} \\ 15{,}795 \end{array}$$

It checks perfectly, doesn't it?

One other checking technique that is sometimes used is comparison of remainders R_{11}. These are a bit more difficult to find than the R_9 but not much more. Any number has the same R_{11} as does the sum of its odd-position digits minus the sum of its even-position digits. The only nuisance, really, is that the difference may be negative, in which case one adds a multiple of 11 to it to make it positive. If you have just finished a very long, very important computation and want to be more certain of your answer than you can be by using R_9 alone, you can use both checks together, which incidentally is equivalent to checking by comparing remainders R_{99}.

EXERCISES

1. Find

   $R_9(17{,}316)$ $\qquad$ $R_9(123{,}456{,}789)$ $\qquad$ $R_9(111{,}111)$

2. Check the following additions by casting out 9's:

1487	389	1007	25
3564	110	3210	37
			97
			13

3. Check the following subtractions by casting out 9's:

$$\begin{array}{r}3564\\-1487\\\hline\end{array} \qquad \begin{array}{r}389\\-110\\\hline\end{array} \qquad \begin{array}{r}3210\\-1007\\\hline\end{array}$$

4. Check the following multiplications by casting out 9's:

$$\begin{array}{r}3564\\\times 1487\\\hline\end{array} \qquad \begin{array}{r}389\\\times 110\\\hline\end{array} \qquad \begin{array}{r}3210\\\times 1007\\\hline\end{array}$$

5. Check the following divisions-with-remainder by casting out 9's:

$$378\overline{)175632} \qquad 56\overline{)13849} \qquad 99\overline{)111111}$$

6. For the case $n = 13$, illustrate Equations (i), (ii), and (iii) with values of a, a_1, a_2, a_3, and b of your own choosing.

7. Describe how to find quickly R_2, R_4, R_5, and R_{10} of any number. Why are these remainders not practical for checking arithmetic problems done in base ten?

8. Find

$R_{11}(17,316) \qquad R_{11}(123,456,789) \qquad R_{11}(111,111)$

9. Prove that a mispositioned multiplication such as the one given in the text can never be detected by casting out 9's. Can it be detected by casting out 11's?

10. Develop a method for finding R_3 of a natural number.

11. Develop a method for finding R_{99} of a natural number.

12. Give proofs of Equations (ii) and (iii) of Theorem 1. (By doing so you should become better acquainted with the proof of Equation (i), since you can model your proofs after it.)

CHAPTER SEVEN

Topics Having to Do with the Number Two

1. Binary Notation

One of the truly great advances in the handling of numbers, especially large ones, was the adoption of what is called Hindu, Arabic, or decimal notation. This notation uses a clever combination of three devices to represent any number, no matter how large or small. The three devices are ten basic symbols known as *digits:* 0, 1, 2, 3, 4, 5, 6, 7, 8, 9, a *decimal point,* and *position* of the digits to the left and right of the decimal point. (If the number represented is a whole number, so that there are no digits to the right of the decimal point, it has been agreed that the decimal point may be omitted, since no confusion will result in this one case.) Some simple examples of numbers written in decimal notation are:

$$123 \quad 3.5 \quad 1704 \quad .796 \quad .0001 \quad .333\cdots.$$

The great advantage of the decimal system over the previously used "Roman numerals" is the speed with which addition, subtraction, multiplication, and division can be performed. The speed comes from the ease with which one can "carry" the overflow from the units position to the tens position, from the tens position to the hundreds position, and so on. What makes carrying so easy is the combined use of position and digits.

1. BINARY NOTATION

The question often arises: Why are there ten digits in the decimal system? Why not twelve, or six, or twenty, or some other number? The answer, it turns out, is no more than three feet away from you, no matter where you are. The answer is in your ten fingers (or *digiti*, as the Romans called them). Men all over the world used to count on their fingers.[1] When they ran out of fingers, they needed to store the count of ten somewhere, so that they would have their fingers free to continue counting. The Hindu method of storing tens, hundreds, and so on by position and using the same ten symbols over again in each position turned out to be the most practical of a great many systems tried by a great many peoples.

If men all had some number n of fingers other than ten, the basic Hindu idea of using symbols and position would still have worked. Instead of ten symbols, the system would have used n of them, whatever n might have been. We call such a system a *base n* system. Our decimal system, then, is the *base ten* system. One of the favorite occupations of teachers of "new math" is exposing their students to base n systems, with, say, $n = 7$ or 8. You may have heard Tom Lehrer sing "New Math," in which he performs some base eight arithmetic to piano accompaniment.

Since arithmetic is hard enough to learn in base ten, one may wonder if it is worth the effort to repeat the struggle with other bases—especially since it does not appear that we will be changing the number of our fingers in the near future. Really, it is not worth the effort to become proficient in more than one system, but it is worthwhile to become mildly acquainted with the one other system in common use today. That system is the base two, or *binary* system, which is used by practically every digital computer in the world today.

The binary system, with its two digits 0 and 1, is the perfect system for a computer to calculate in, because all arithmetic in a computer is done by means of switches (usually semiconductor devices like transistors), and a simple switch has just two states to be in: open (nonconducting) and closed (conducting). In view of the tremendous speed of these machines (addition, for instance, performed in one-millionth of a second) and therefore the immense amount of computations they perform, we can say that the binary system today is more commonly used than the decimal system.

Another reason for studying the binary system is that the binary system can occasionally be used naturally in theory, while the

---

[1] Some early peoples counted on both fingers and toes and accordingly used a base twenty system. The English word *score* and the French words *vingt* (20) and quatre-vingt (80) are remnants of base twenty systems.

decimal system cannot. This is due to the fact that the number two often occupies a special position in mathematics. We will see an example of a natural occurrence of the binary system in the next section.

Right now, let us develop methods for converting from binary notation to decimal notation and vice versa. For simplicity we will consider only natural numbers.

Binary to Decimal

Let us convert the binary number 1001101 to decimal form. To make the conversion, all we need to know is which of the binary positions (units, twos, fours, eights, and so on) contain the digit 1:

sixty-fours	thirty-twos	sixteens	eights	fours	twos	units
1	0	0	1	1	0	1

For our number those positions are: sixty-fours, eights, fours, and units. Therefore, in decimal notation, our number is

$$64 + 8 + 4 + 1 = 77.$$

Decimal to Binary

The reverse conversion is trickier to discover or remember, but it is just as easy to carry out. Pretend for the moment we do not know the binary representation of decimal 77.

Decimal 77 will have a 1 in the binary units position due to the fact that decimal 77 is an odd number. That is, 77 leaves a remainder of 1 when divided by 2:

$$\begin{array}{r} 38 \quad 1 \\ 2\overline{)77} \end{array}$$

Now think of decimal 77 as $2 \cdot 38 + 1$. Decimal 77 will have a 0 in the binary twos position due to the fact that 38 is an even number. That is, 38 leaves a remainder of 0 when divided by 2:

$$\begin{array}{r} 19 \quad 0 \\ 2\overline{)38} \end{array}$$

Now think of decimal 77 as $4 \cdot 19 + 1$. Decimal 77 will have a 1 in the binary fours position due to the fact that 19 is an odd number. That is, 19 leaves a remainder of 1 when divided by 2:

$$\begin{array}{r} 9 \quad 1 \\ 2\overline{)19} \end{array}$$

So far, we have found the binary units digit to be 1, the binary twos digit to be 0, and the binary fours digit to be 1. The process of

1. BINARY NOTATION

dividing each quotient in succession produces as remainders the various binary digits, in succession. Now that we know this, it is easy to arrange the divisions in a neat stack and simply watch the binary representation build up, *with the units digit on the bottom*:

$$
\begin{array}{r|l}
 & 0 \quad 1 \\
2) & \overline{1} \quad 0 \\
2) & \overline{2} \quad 0 \\
2) & \overline{4} \quad 1 \\
2) & \overline{9} \quad 1 \\
2) & \overline{19} \quad 0 \\
2) & \overline{38} \quad 1 \\
2) & \overline{77} \\
\end{array}
$$

EXERCISES

1. Find the decimal representation of the following numbers written in binary notation: 100000, 11111, 1001011, 11001000, 1010.

2. Find the binary representation of the following numbers written in decimal notation: 45, 30, 79, 1111.

3. The so-called perfect numbers 6, 28, 496 were discussed in Chapter 2. Write each of these numbers in binary notation. Do you see a pattern?

4. Write the first thirty-two natural numbers (1, 2, 3, · · · , 31, 32 in decimal notation) in binary notation. [*Hint:* Obtain each successive number by adding 1 to the preceding number using base two addition. That should be easier than performing a large number of divisions, which you would need if you used the method employed in Exercise 2.]

5. Find 10011 + 1101 by adding in the binary system. (Use the same addition technique you learned for the decimal system, except "carry" when you add two 1's, because in the binary system 1 + 1 = 10.) Check your answer by converting all numbers to decimal form.

6. Find 10011 − 1101 by subtracting in the binary system. Check your answer by converting all numbers to decimal form.

7. Recheck your work in Exercises 5 and 6 by subtracting and adding (respectively) in the binary system.

8. Find 10011 × 1101 by multiplying in the binary system. [*Hint:* You will probably find it easier to add the partial products two

at a time, rather than all three at once.] Check your answer by converting all numbers to decimal form.

9. Recheck your work in Exercise 8 by dividing in the binary system.

10. Repeat Exercises 5 through 9 for the pairs of numbers (in binary notation) 110001 and 111.

11. If a natural number n is divided by the number four, it must leave one of the remainders zero, one, two or three. If n is written in binary notation, describe how to tell by sight what the remainder will be.

12. Applying Exercise 11 to the natural number $n = 1101110010$ 1000011010 (binary), tell what the remainder is after division by four. What is the quotient? Keep your answers in binary notation.

13. Look up the meanings of the words "gill" and "pottle," used as units of measure. Explain how gills, cups, pints, quarts, pottles, and gallons can be put together into a binary system for measuring liquids. (These units of measure, plus more along the same lines, like pecks, firkins, and hogsheads, originated in England some 700 years ago.)

14. Show that if a number is d digits long in its decimal representation, then it is at most $4d$ digits long in its binary representation. [*Hint:* In the division technique for converting from decimal to binary, show that every fourth dividend will be at least one digit shorter in length.] Give an example of a number whose binary representation has four times the digits of its decimal representation.

15. What technique in the binary system corresponds to casting out nines in the decimal system? Explain why this technique has no practical value.

16. What technique in the binary system corresponds to comparing remainders R_{99} in the decimal system? Describe how to make the check in the binary system, using the addition and multiplication from Exercises 5 and 8 as examples. Are there many errors to which this test is not sensitive?

2. Nim

The origins of the game of Nim are lost in antiquity—so lost that no one knows whether "antiquity" means the nineteenth century or many millennia ago. Nim is a game played between two persons, with piles of sticks laid out at the start of the game. The rules are:

(1) On each turn, the player whose turn it is must pick up at least one stick, but he can pick up as many sticks as he wishes, all from one pile.

(2) The person picking up the last stick *loses*.

[Rule (2) can be reversed without altering the essence or strategy of the game.]

Nim has been popular off and on for at least one century. Recently, it became very popular after its appearance in the film *Last Year at Marienbad*. Suppose in playing the game you had some strategy by which you could force your opponent to leave you the last pile containing more than one stick. You would then have *complete control* over who takes the last stick. It turns out that there is such a strategy, provided you can once maneuver into what is called a "safe" position. This strategy, based on the binary system, was discovered around the turn of the century by C. L. Bouton (1869–1922), a professor at Harvard University.

In describing this strategy, we will take for an example the traditional starting configuration of the game—four piles of sticks, containing 1, 3, 5, and 7 sticks. The first step is to write down in binary notation the number of sticks in each pile, one number under the next:

$$\begin{array}{ll} 111 & (7 \text{ sticks}) \\ 101 & (5 \text{ sticks}) \\ 11 & (3 \text{ sticks}) \\ 1 & (1 \text{ stick}) \end{array}$$

Step two is to sum each column as though the entries were integers:

$$\begin{array}{r} 111 \\ 101 \\ 11 \\ \underline{1} \\ 224 \end{array}$$

After each move in the game, you will need to repeat these two steps. With a little practice you will be able to do them in your head. Notice that in the above sum each column of 0's and 1's adds up to an even number. Any configuration of sticks for which every column sum is even is called a *safe position*. Any other configuration is called an *unsafe position*. The reason for these names is that if one player once manages to take sticks so as to achieve a safe position, then that same player will always be able to achieve safe positions in the future, and if he continues to do so, he will have control of the game. The following theorem contains the information we need to see this.

THEOREM 1. Any move starting from a safe position will produce an unsafe position. Given an unsafe position, there always exists a move that will produce a safe position.

Proof. Suppose you start from a safe position. In the sum described above, locate the row corresponding to the pile from which you decide to remove sticks. At least one binary digit in that row is altered by removal of the sticks, simply because the number of sticks in the pile changes. No other row is altered, since sticks may be taken from only one pile. Therefore, there will be at least one column in which exactly one entry changes, and so that column sum is changed from even to odd. Therefore, the position produced is unsafe.

Now suppose you start from an unsafe position. Find the leftmost column which sums to an odd number, and then find a row which contributes a 1 to that sum. Remove a suitable number of sticks from the corresponding pile so as to change that row to one which makes every column sum even. This is possible, since you may leave behind any number of sticks less than the original number. This completes the proof.

Now let us see what happens if you follow the strategy of always returning to safe positions. (By the theorem, you can do this if your opponent even once produces an unsafe position.) Is it possible that your opponent will be left with the last multiple-stick pile (and consequently complete control)? No, because such a configuration will have at least one column sum odd, so you would not have produced a safe position that time. Therefore, if it is not your opponent who is left with the last multiple-stick pile, it must be *you* who is left with it. At the time you are left with it, *you* will decide who is to pick up the last stick.

If you want your opponent to take the last stick, you should leave an *odd* number of single-stick piles. (It is important for you to realize that in this moment of decision, you are *not* producing what is defined to be a safe position. Leaving a safe position at this point will force *you* to take the last stick. Just a warning in case you are prone to doing too much of a good thing.) On the other hand, if you are playing under the rule that the person who picks up the last stick wins, then you will want either to leave an even number of single-stick piles or else pick up all the sticks in that one last pile.

The best way to get the strategy down pat is to go out and play a few games. Skill comes fast; in fact, you probably will find that after a very few times you can do all the computations in your head. For illustration and practice, let us do one example here. We will start with the traditional 1-3-5-7 configuration, although, of course, the strategy applies to any number of sticks in any number of piles. Notice that 1-3-5-7 is a safe position, so if we want to be absolutely certain of winning, we should let our opponents go first.

Figure 1 illustrates the entire sequence of moves leading to our win. Our opponents first choose to remove 3 sticks from the 5 pile. This causes an odd sum in the left-most column, and to change that and all other sums back to even, we must remove sticks from the 7 pile. It turns out that we must remove all 7 of them, so the configuration after two moves is 1-3-2. Our opponents elect to remove the 1-stick pile. This causes one odd sum, in the right-most column, and to produce a safe position we now must remove sticks from a pile contributing to that sum. Removing one stick from the 3 pile returns a safe position. Our opponents are now in a position that any move they make will leave us with the last multiple-stick pile. They remove a single stick from one pile. With the last multiple-stick pile at our fingertips, we have control. We pick up that entire pile, leaving the very last stick for them, and so we win.

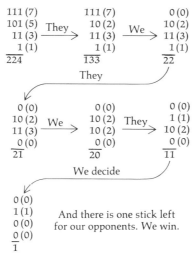

FIGURE 1. Example of the strategy of maintaining safe positions in the game of Nim.

EXERCISES

1. Play two games of Nim with someone, letting him start with the standard 1-3-5-7 position. Win both games. Write down a record of each game similar to Figure 1.

2. Play two games of Nim with someone, as in Exercise 1, except this time you start with the standard 1-3-5-7 position. Try to win both games. (If he fails to produce a safe position after your first move, you should produce one after his move. After that, all you need to do is follow the usual strategy.) Keep a record of each game.

3. Suppose you are playing a game of Nim with just two piles of sticks. What are the safe positions?

4. In which of the following Nim games would you wish to have the first turn, and in which would you wish your opponent to have the first turn, so that you would always be certain of winning?

   (a) 2-4-6-8
   (b) 1-5-9-13
   (c) 1-2-3-5-7-11-13
   (d) 2-4-8-16-32
   (e) 2-5-7-15-21
   (f) 10-20-30-40

5. In the games of Exercise 4 in which you should start first to be certain of winning, tell what your first move should be in each case.

6. In Exercise 4, is there any case in which you can reach a safe position by two different opening moves?

7. Judging by your answer to Exercise 4, would you guess that safe or unsafe positions are the more common configurations in Nim? Write down at random ten different starting positions, then see how many of these are safe.

8. How many members of your class obtained five or more safe starting positions out of the ten positions they wrote down in Exercise 7?

9. For the unsafe starting positions you wrote down in Exercise 7, tell how to achieve a safe position from each one. Are there any cases in which it is possible to achieve a safe position in more than one way?

10. Alter Rule 1 of Nim to read: On each turn, the player whose turn it is must pick up at least one stick, but he can pick up as many sticks as he wishes from at most two piles. Then redefine *safe position* to be any configuration of sticks for which every column sum is divisible by 3. Prove Theorem 1 in this case.

11. Play two games of the Exercise 10 version of Nim with someone. Try to win both games. Write down a record of each game similar to Figure 1.

12. Alter Rule 1 of Nim to read: On each turn, the player whose turn it is must pick up at least one stick, but he can pick up as many sticks as he wishes from at most k piles. How would you redefine *safe position* in this case, in order that Theorem 1 should be true?

3. The Tower of Brahma

" 'Tis said that in the great temple at Benares, beneath the dome which marks the center of the world, fixed into a slab of brass there are three diamond needles, each a cubit high and as thick as the body of a bee. On one of these needles, at the beginning of the ages, God placed sixty-four discs of pure gold, the largest resting on the brass slab and the others decreasing in size to the top. That is the Tower of Brahma. Night and day the priests take turns transferring the discs from needle to needle, without deviating from the fixed and immutable laws imposed by Brahma: The priest may move only one disc at a time, and he must place this disc only on a free needle or on top of a larger disc. When in strict accordance with these rules the sixty-four discs have been transferred from the needle on which God placed them to one of the other needles, tower and Brahmins will crumble to dust and the world will cease to exist."

In reality, the Tower of Brahma (or Tower of Hanoi) described in the above quote was invented by the number-theorist Edouard Lucas in 1883. The description given above was published tongue-in-cheek by a friend of his one year later. The puzzle that one sometimes sees in toy stores has a much smaller number of discs, because with 64 discs, it would take many more years than the earth is old to complete the transfer "without deviating from the fixed and immutable laws imposed by Brahma."

In this section we will develop the method of working the puzzle in the least number of steps for any arbitrary number n of discs. As a step toward developing the method, let us first look at the smallest cases, to see if we can find a pattern that will carry over to the general case. Always suppose we are starting with the puzzle as in Figure 2, with the discs stacked in order on Needle A. We will number the discs 1 though n according to size, Disc 1 being the smallest.

In the case $n = 1$, of course, one move does the trick. We do not even need to use the third needle. In case $n = 2$, as shown in Figure 3, we already begin to see a little something of interest. We can do the transfer as follows: Disc 1 to Needle B, Disc 2 to Needle C, and

Disc 1 to Needle C, three moves in all. In the case $n = 3$ (see Figure 4), we can do as follows: Do the three moves of the case $n = 2$. Then move Disc 3 to Needle B. Then move the two discs on Needle C to Needle B; this can be accomplished in three moves similar to those of the case $n = 2$. In all, $3 + 1 + 3 = 7$ moves are used. In the case $n = 4$, we can transfer the tower by performing the 7 moves of the case $n = 3$, then transferring the largest disc, then performing 7 more moves to place the three smaller discs on top of it. In all, $7 + 1 + 7 = 15$ moves are used.

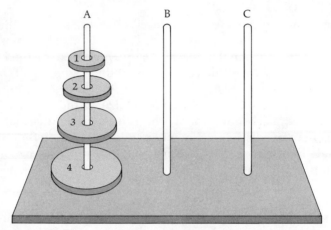

FIGURE 2. The Tower of Brahma (or Hanoi), with four discs.

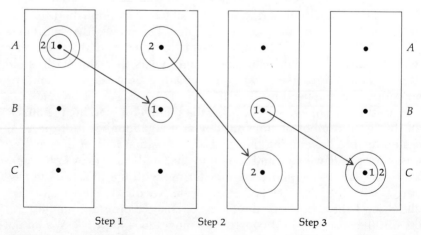

FIGURE 3. The three steps in transferring a Tower of Brahma made of two discs.

3. THE TOWER OF BRAHMA

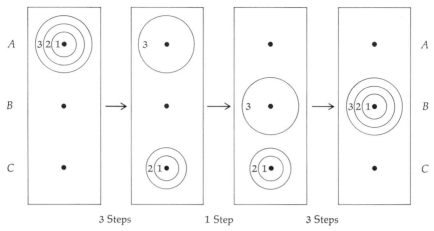

FIGURE 4. The seven steps in transferring a Tower of Brahma of three discs. Three steps take the top two discs off Disc 3 and onto one other needle, then Disc 3 is moved, and finally three more steps move the smaller two discs back onto Disc 3.

Do you see a pattern developing? It involves the number 2 very intimately, for in each of these four cases the number of moves is 1 less than a power of 2. In fact, for $n = 1, 2, 3$, or 4, we have found that the transfer can be accomplished in $2^n - 1$ moves. We will now use the philosophy with which we passed from the case of 2 discs to 3 discs and from 3 discs to 4 discs to prove the following theorem.

THEOREM 2. *In the Tower of Brahma, the minimum number of moves needed to transfer n discs from one needle to another is equal to one more than twice the minimum number of moves needed to transfer $n - 1$ discs from one needle to another.*

Proof. Let us look carefully at the process pictured in Figure 5 of transferring the n discs. Before the largest disc can be moved, the $n - 1$ smaller discs must all be moved off Needle A. Furthermore, the $n - 1$ smaller discs must all be on one other needle, say C, for otherwise the largest disc would have nowhere to go. Therefore, in order to free the largest disc, we must complete a full transfer of $n - 1$ discs. This transfer can be accomplished in the same way as it would be were the largest disc not present, since as long as it is on the very bottom, it imposes no restriction on the movement of the other discs.

Once the largest disc has been freed to move, we use one move to place it on the empty Needle B, and then we proceed to move the

$n-1$ smaller discs onto it. Again, the transfer of the $n-1$ smaller discs can be done in the same way as it would be done if the largest disc were not present.

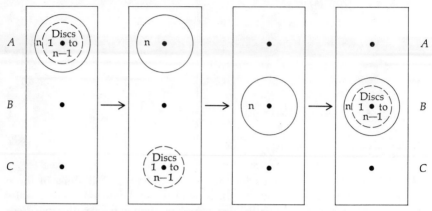

FIGURE 5. The basic idea of the proof of Theorem 2. Before Disc n can be moved, the $n-1$ smaller discs must be transferred to another needle. Then Disc n is moved, and finally the $n-1$ smaller discs are transferred onto it.

We have therefore found the way to transfer the n discs from one needle to another in the minimum number of moves. This minimum number of moves involved two complete transfers of $n-1$ discs plus one more move, as was to be shown. This completes the proof.

THEOREM 3. In the Tower of Brahma, the minimum number of moves needed to transfer n discs from one needle to another is $2^n - 1$.

Proof. In the case of 1 disc, the theorem is true, because $2^1 - 1 = 1$, and exactly one move is required. We will prove the theorem by showing that if it is true for the case of $n-1$ discs, then it is also true for the case of n discs.

Therefore assume that it is true for the case of $n-1$ discs. That is, assume that the minimum number of moves needed to transfer $n-1$ discs is

$$2^{n-1} - 1.$$

By Theorem 2, then, the minimum number of moves needed to transfer n discs is

$$2(2^{n-1} - 1) + 1 = 2^n - 2 + 1$$
$$= 2^n - 1,$$

as was to be shown. This completes the proof.

3. THE TOWER OF BRAHMA

The proof of Theorem 3 is a good example of the technique known as *mathematical induction*. Proving that the truth of the theorem for case $n-1$ implies the truth of the theorem for case n is like standing dominoes in a line so that each one will knock the next one down if it falls into it. Proving that the theorem holds for the case $n=1$ is like knocking over the first domino; when it falls, all the rest fall as well.

If you are interested in working a small Tower of Brahma puzzle, notice that Theorems 2 and 3 not only tell how many moves must be made, but also *what* moves must be made. To transfer 10 discs, it is necessary to complete two transfers of the 9 smaller discs, with one move in between. To do each transfer of the 9 smaller discs, it is necessary to complete two transfers of 8 discs, with one move in between. And so on.

While the solution of the puzzle in general is interesting, it begins to look as though the actual working of any one puzzle could be very repetitious and boring. You might like to try working out the case $n=5$ (or 4 or 6) with playing cards, just to see if there is any challenge. For $n=5$, 31 moves are the minimum number necessary. If you find it easy, however, try giving the puzzle to someone who has not seen the general method worked out. To him it will be considerably more challenging.

EXERCISES

1. Write out the seven steps needed to transfer three discs from Needle A to Needle C. [*Suggestion:* Describe each step by telling which discs are on which needles when that step is completed.]

2. Repeat the work of Exercise 1 for the case of 4 discs. [*Hint:* For 4 discs, remember that almost half of the work is done in Exercise 1.]

3. The numbers below represent 5 discs on 3 needles during the process of transferring the Tower of Brahma from Needle A to Needle B. What should the next move be? How many moves have been performed so far? How many moves are there left to perform?

    $$\begin{array}{ccc} 5 & 1, 2, 3 & 4 \\ A & B & C \end{array}$$

4. Estimate (roughly) how long it would take you to move a single disc in the Tower of Brahma puzzle. Then for a tower of 15

discs, calculate how long it would take you to transfer the entire tower from one needle to another. Do the same for a tower of 25 discs.

5. Few people have a clear idea of how fast the number 2^n grows as n increases. For that reason they find the number of moves required to transfer the 64-disc Tower of Brahma simply incredible. Yet many natural phenomena exhibit that same rapid, "exponential" growth, so understanding it is important. Write out in decimal form the number of moves necessary to transfer the 64-disc Tower of Brahma from one needle to another. Write this same number in binary form.

6. According to legend, the inventor of the game of chess requested as his reward one grain of wheat for the first square of the chessboard, two for the second, four for the third, eight for the fourth, and so on. A chessboard contains 64 squares. Write out in binary form the number of grains of wheat that he requested. Write the same number in decimal form. Assuming there are 7000 grains to the pound, how many tons of wheat did he request?

7. A sheet of looseleaf paper is approximately 1/250 of an inch thick. The moon is approximately 240,000 miles from the earth. Show that if a sufficiently large sheet of looseleaf paper were folded (doubled) over on itself 42 times, the resulting wad would be thicker than the distance from the earth to the moon.

8. How many times can you fold a standard size sheet of looseleaf paper before it becomes too thick to fold? How thick is the wad you end up with? (One of my students once did the equivalent of 8 folds, doing the last few by cutting when the wad got too thick to fold. She presented me with a neatly stacked and taped wad slightly over one inch thick.)

4. The Fifteen Puzzle

Many puzzle collectors consider Sam Loyd to be the greatest inventor of puzzles who ever lived. If you are not a buff, you may never have heard of Sam Loyd, but you surely have seen his most famous puzzle. It consists of 15 square blocks, usually numbered 1 to 15, placed in a shallow square tray just large enough to hold 16 blocks, as in Figure 6.

The blocks and tray are tongue-and-grooved so that the blocks will slide freely within the tray but cannot be lifted out. When sold,

4. THE FIFTEEN PUZZLE 143

the puzzle usually is accompanied by a list of interesting configurations into which the blocks might be slid, some of them labeled "possible" and others labeled "impossible." In this section we will discover the reason why some configurations really are impossible to achieve by sliding the blocks. That reason has a lot to do with even and odd numbers.

Over the years there have been published a number of unnecessarily complicated explanations of the puzzle. I confess that I myself once published one of those overly complicated accounts. The tendency to overcomplicate seems to stem from the desire to consider the blank space as a special entity, all by itself. Study of the puzzle's behavior becomes much simpler if we agree to treat the blank space much like any other block (say "block 16") and then watch its progress through the puzzle as the blocks are slid in various directions.

1	2	3	4
5	6	7	8
9	10	11	12
13	14	15	

FIGURE 6. Sam Loyd's Fifteen Puzzle. The fifteen blocks are free to slide within the tray, but they cannot be removed.

By a "move" we will always mean the act of sliding a block next to the empty space *into* the empty space. In terms of block 16, a move consists of switching the positions of block 16 and some block next to it, as in Figure 7.

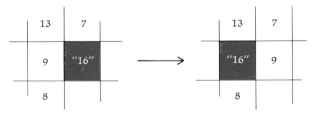

FIGURE 7. A typical "move" in the Fifteen Puzzle. Block 9, next to the empty space, is slid into the empty space. For purposes of studying the puzzle, it is helpful to think of this move as a switch of the two blocks, 9 and "16."

Now suppose we begin with all the blocks of the puzzle in numerical order, as they are in Figure 6. We imagine that beneath the blocks, inscribed on the tray bottom, is a checkerboard pattern of E's and O's as in Figure 8. In the beginning, block 16 is in an E space. After the first move, it will be in an O space. After the second move, it will be in an E space again. With each succeeding move, the type of space, E or O, in which block 16 is located must change, since above, below, right, and left of each O space is either a wall of the tray or an E space, and above, below, right, and left of each E space is either a wall of the tray or an O space, as in Figure 9.

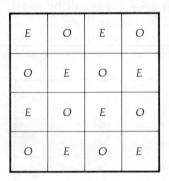

FIGURE 8. Checkerboard pattern of E's and O's imagined to be inscribed in the tray bottom of the puzzle. The pattern is used to tell whether starting with Figure 6 an even or an odd number of moves is needed to reach a given configuration.

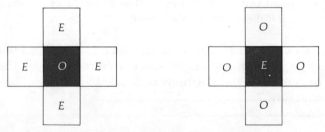

FIGURE 9. If block 16 is in an O space, it must move (up, down, right, or left) to an E space; if block 16 is in an E space, it must move to an O space.

Therefore, no matter how the moves are performed, after an odd number of moves beginning with Figure 6, the empty space *must* be an O space, and after an even number of moves beginning with Figure 6, the empty space *must* be an E space. Now you can see why we used the letters E and O to label the spaces.

Suppose you begin with the puzzle as in Figure 6 and try to wind up with a configuration in which the blank space (block 16) is an E

space. Then you know that if the configuration is to be achieved, it must be accomplished through an even number of moves. Now it turns out that there are some rearrangements of the numbers 1 through 16 that can *never* be achieved by an even number of interchanges of two numbers at a time. (By rearrangement of the numbers 1 through 16, we mean any ordering of them, not necessarily obtained by sliding the blocks within the puzzle.) If the configuration you are trying to reach is of this type, it follows from block 16 being in an E space that you will not be able to achieve it by sliding the blocks.

Similarly, if the desired configuration is to have block 16 in an O space, you would have to reach it through an odd number of moves. But there are some rearrangements of the numbers 1 through 16 that can never be achieved through an odd number of interchanges of two numbers at a time. If the configuration you are trying to reach is of this type, then because block 16 is in an O space, you will not be able to achieve it by sliding the blocks.

The definition and theorem below tell the full story about the possibility of obtaining various rearrangements of the numbers 1 through n by a series of interchanges of two numbers at a time.

DEFINITION. The interchange of a pair of numbers is called a transposition.

THEOREM 4. Any rearrangement of the numbers 1 through n can be accomplished by some series of transpositions. If a given rearrangement of the numbers 1 through n can be achieved by an even number of transpositions, then every series of transpositions which produces that rearrangement will consist of an even number of transpositions. If a given rearrangement can be achieved by an odd number of transpositions, then every series of transpositions which produces that rearrangement will consist of an odd number of transpositions.

We will postpone most of the proof of Theorem 4 to the next section. You can see what an amazing story the theorem tells. By means of it, we can divide the collection of all possible rearrangements of the numbers 1 through n into two classes: the *even* rearrangements (those obtainable by an even number of transpositions) and the *odd* rearrangements (those obtainable by an odd number of transpositions). By Theorem 4, each possible rearrangement is either even or odd and *cannot* be both at once.

Now let us develop a technique for determining whether a given rearrangement is even or odd. Such a technique, of course, is what we need if we are to apply Theorem 4 to the Fifteen Puzzle. It will, incidentally, furnish us with a proof of the first sentence of Theorem 4.

146   CHAP. 7   TOPICS HAVING TO DO WITH THE NUMBER TWO

What we do is work backwards; starting from a given rearrangement, we apply transposition after transposition until the rearrangement is converted to the normal ordering of the numbers 1 through n. The same series of transpositions applied in reverse order will convert the normal ordering into the given rearrangement. One concrete example will suffice to illustrate the technique.

Let us suppose we are given the following rearrangement of the numbers 1 through 9:

$$5 \quad 2 \quad 4 \quad 7 \quad 3 \quad 9 \quad 1 \quad 8 \quad 6$$

To put the number 1 into its proper position, we can switch it in succession with the numbers 9, 3, 7, 4, 2, 5—all the numbers that are to its left and greater than it. Then to put the number 2 into its proper position, we can switch it with the number 5—the only number to its left and greater than it. Then to put the number 3 into its proper position, we can switch it in succession with the numbers 7, 4, 5—all the numbers that are to its left and greater than it.

Notice that in each case it is possible to tell from the original rearrangement how many switches should be made to put a number like 2 or 3 into its proper position. The reason is that the earlier switches only move smaller numbers to the left of larger ones. Continuing, as in Figure 10, we see that we can put the remaining 6 numbers into their proper positions by 1, 0, 3, 0, 1, and 0 switches, respectively. Altogether, to convert the given rearrangement to the normal ordering of the numbers 1 through 9,

$$1 \quad 2 \quad 3 \quad 4 \quad 5 \quad 6 \quad 7 \quad 8 \quad 9,$$

we have used

$$6 + 1 + 3 + 1 + 0 + 3 + 0 + 1 + 0 = 15$$

transpositions. Therefore, the given rearrangement is an odd rearrangement.

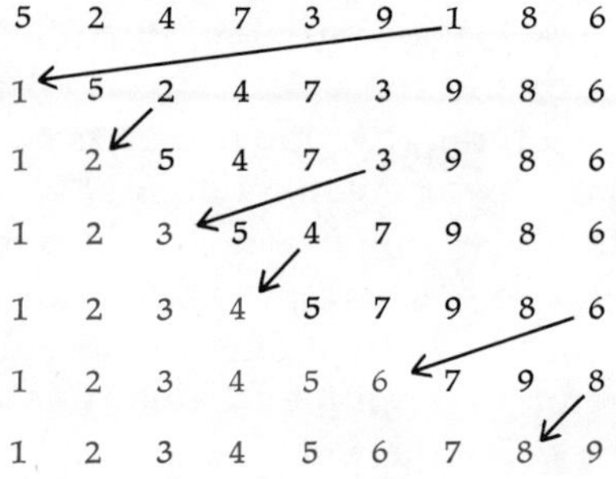

FIGURE 10. Converting the rearrangement 5, 2, 4, 7, 3, 9, 1, 8, 6 to the normal order 1, 2, 3, 4, 5, 6, 7, 8, 9 by a series of fifteen transpositions.

4. THE FIFTEEN PUZZLE

In general, then, to determine evenness or oddness of a rearrangement, all we need to do is count the number of times a larger number appears to the left of a smaller number. In the above example this number was 15. If the result of the count is even, the rearrangement is even; if the result of the count is odd, the rearrangement is odd.

The rearrangement of the puzzle shown in Figure 11 is an example of an odd rearrangment, since it can be obtained from Figure 6 simply by trading the numbers 14 and 15 — that is, by a single transposition. (Equivalently, only *once* in the rearrangement does a larger number precede a smaller number.) Since the empty space is an E space, it follows that the configuration of Figure 11 cannot be obtained from that of Figure 6 by sliding the blocks. Sometimes the puzzle has been manufactured in the configuration of Figure 11. Since Figure 6 cannot be slid into Figure 11, neither can Figure 11 be slid into Figure 6 — and that is frustrating! Sam Loyd offered one thousand dollars to anyone who could produce Figure 6 from Figure 11; naturally, he never had to pay the money. At the time he made his offer, it had not yet been proved that that little task was impossible. However, he must have been very confident that no one would find a way of doing it.

1	2	3	4
5	6	7	8
9	10	11	12
13	15	14	

FIGURE 11. The Fifteen Puzzle with two blocks interchanged from their normal order. This configuration cannot be obtained from Figure 6, and Figure 6 cannot be obtained from it.

Another version of the puzzle that is very clever—almost fiendishly so—is the letter puzzle shown in Figure 12. The top 8 blocks are one color, and the bottom 7 are another. Notice that there are two *R*'s of the same color. You can show this puzzle to a friend as it appears in Figure 12, scramble it in such a way that the *R* of YOUR winds up in the upper left-hand corner, and then challenge him to put the puzzle back in order. Chances are that he will leave the *R* where it is as he proceeds to put the other blocks back in position. What he will be trying to do then is achieve an odd rearrangement of Figure 12 with the blank space an *E* space. He simply will not succeed. He will be able to spell out *RATE YOUR MIND PLA*, but he will not be able to interchange the *L* and the *A* unless he also interchanges the two *R*'s way above.

All this time we have been discussing configurations that Theorem 4 says cannot be achieved. Of those configurations not ruled out by Theorem 4, which ones can be achieved? It turns out that any rearrangement not ruled out by Theorem 4 can actually be achieved by sliding the blocks. That is, starting with Figure 6, any even rearrangement with the blank space an *E* space can be attained, and any odd rearrangement with the blank space an *O* space can be attained. To prove this is somewhat involved, so we will not attempt it here. There are a number of proofs known, but they all get bogged down in sticky details.

R	A	T	E
Y	O	U	R
M	I	N	D
P	A	L	■

FIGURE 12. A puzzle to give to friends. Notice the two *R*'s. Interchanging them makes the puzzle unworkable.

EXERCISES

1. Which of the following rearrangements of the numbers 1, 2, 3, 4, 5, 6 are even and which are odd?

   2, 3, 4, 5, 6, 1   6, 5, 4, 3, 2, 1   2, 1, 4, 3, 6, 5
   4, 5, 6, 3, 2, 1   3, 6, 2, 5, 1, 4

2. How many of the rearrangements of the numbers 1, 2, 3 are there? List them all. Tell which rearrangements are even and which are odd.

3. In the *RATE YOUR MIND PAL* puzzle, switching the two *R*'s is a single transposition and thus produces an odd rearrangement of the puzzle. However, we can also use the method developed

4. THE FIFTEEN PUZZLE   149

in the text for determining evenness or oddness: If the puzzle had been numbered 1 through 16 instead of lettered, switching the same two blocks would produce the rearrangement

8 2 3 4 5 6 7 1 9 10 11 12 13 14 15 16.

In this rearrangement, count the number of times a larger number appears to the left of a smaller number. Is the number of times even or odd?

4. In the *RATE YOUR MIND PAL* puzzle, if you switch the two R's and your friend working the puzzle switches the two A's, can he then spell out *RATE YOUR MIND PAL*? Why? Why is it not very likely that he would switch the two A's?

5. Tell whether the puzzle of Figure 6 can be put into the following configurations by sliding the blocks. (The number 16 represents the blank space.)

1	2	3	4		16	15	14	13		1	2	3	4
12	13	14	5		12	11	10	9		8	7	6	5
11	16	15	6		8	7	6	5		9	10	11	12
10	9	8	7		4	3	2	1		16	15	14	13

6. The "three" puzzle shown below can also be arranged in various configurations by sliding the blocks. There are 12 configurations that can be achieved, 3 of them for each of the 4 positions of the empty space. List all 12 configurations.

7. For each rearrangement in Exercise 1, write down a series of transpositions which produces that rearrangement. [*Hint:* You can do this either by guess or by the method developed in the text. When you use the method of the text, remember that the order of the transpositions must be reversed to produce the rearrangement from the normal order.]

8. Repeat Exercise 2 for the four numbers 1, 2, 3, 4. [*Hint:* So that you will not feel lost, there are exactly 24 rearrangements. Now try to develop a scheme that will enable you to write them all down in some sort of order. For example, you could first write down the 6 rearrangements in which 1 is replaced by 1, then the 6 in which 1 is replaced by 2, and so on.]

9. (Open question.) Try to think of a more natural-sounding challenge than *RATE YOUR MIND PAL* that can be used in the same way. Feel free to arrange the coloring differently if that will help.

5. Establishing the Difference between Even and Odd Rearrangements

This section is devoted to a proof of the main part of Theorem 4. Since we will present nothing here but the proof, you may find this section somewhat heavier going than the rest of the book. It is included more for the sake of completeness than beauty, although I believe the proof here to be one of the nicest of several hundred known proofs of Theorem 4.

To begin, we will adopt the notation (a, b) for the transposition which switches the two numbers a and b, a being different from b. Notice that (b, a) represents the same transposition as (a, b) does. When we write down a series of transpositions

$$(a, b)(c, d)(e, f) \cdots,$$

we will understand that to mean the rearrangement produced by interchanging a and b first, then c and d, then e and f, and so on.

Now if a, b, c, and d are all different numbers, then it is clear that

$$(c, d)(a, b) = (a, b)(c, d). \tag{1}$$

If a, b, and c are all different numbers, then it turns out that

$$(b, c)(a, b) = (a, c)(b, c) \tag{2}$$

and

$$(a, b)(a, c) = (a, c)(b, c). \tag{3}$$

Proof of (2) and (3) consists in seeing what effect each side of the equation has on the three numbers a, b, and c.

Now suppose that the second or third sentence in Theorem 4 is false. In either case, it must then be possible to obtain some rearrangement of the numbers 1 through n both by an even number and by an odd number of transpositions. Apply the even number of transpositions to the numbers 1 through n in their natural order. Then to the result apply the odd number of transpositions *in reverse order*. The end result is the numbers 1 through n back in their natural order, obtained by applying an odd number of transpositions.

That is, there is some series containing an odd number of transpositions which has no net effect on the ordering of the numbers 1 through n. Among all such series of odd length, pick one with the *smallest* (odd) number of transpositions. Suppose it begins with the number a:

$$(a, \ ) \ (\ , \ ) \ \cdots .$$

Among all such minimal odd-length series beginning with a, pick one containing the least number of a's. We will derive a contradiction from the combination of these two minimal properties of the series.

First of all, we notice that somewhere along the series there is at least one other transposition containing the number a, for otherwise the effect of the whole series of transpositions would be to move a out of its original position.

Now, by applying (1) and (2), whichever is appropriate at each stage, we can move the second transposition containing a to the left, until it becomes the second transposition of the series. This is done without changing the number of transpositions in the series and without increasing the number of a's. We therefore wind up with a series of transpositions satisfying both minimality conditions and which in addition looks like

$$(a, b) \, (a, b) \, \cdots$$

or

$$(a, b) \, (a, c) \, \cdots ,$$

where c is different from a and b.

In the first case, the two transpositions cancel each other, and so we have a shorter odd-length series of transpositions which does not move any of the numbers 1 through n. In the second case we can apply (3) and obtain a series of transpositions of the same minimum length, but with a smaller number of a's. In either case we have arrived at a contradiction, so both the second and third sentences of Theorem 4 must be true. This completes the proof.

EXERCISE

1. Prove that

$$(b, c) \, (a, b) \qquad (a, c) \, (b, c) \qquad (a, b) \, (a, c)$$

all have the effect of replacing the number a with b, the number b with c, and the number c with a. Conclude, then, that Equations (2) and (3) are both correct.

CHAPTER EIGHT

Matrices

1. Introduction to Matrices and Their Arithmetic

Imagine that we have just transported ourselves to the local supermarket. We are watching a man on his way home from work picking up a few items his wife needs for the evening meal. His shopping list reads:

> 2 quarts skim milk
> 3 candles
> 1 box napkins
> 10 hard rolls
> 2 frozen dinners.

He finds, as he picks up the items, that the prices per unit item are respectively

$.32 $.20 $.41 $.05 $.63.

How much money (not counting sales tax) does he spend? To find the answer, we line up each item's price opposite the number of items of that kind, multiply, and then add, as in Figure 1. We find that he spends a total of $3.41, not counting tax.

1. INTRODUCTION TO MATRICES

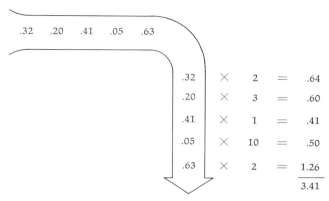

FIGURE 1. Multiplying a price list times a shopping list, to find the total amount of money spent. Such list multiplications occur in very many applications of mathematics.

Suppose, now, we return one week later and find him with much the same list, the only things changed being the numbers on the list:

> 1 quart skim milk
> 0 candles
> 0 boxes napkins
> 50 hard rolls
> 5 frozen dinners.

How much money (not counting sales tax) does the shopper spend in the two weeks combined?

There are *two* ways in which this question can be answered — both giving the same answer, naturally. The one way is to calculate how much money he spends each week, then add the two amounts of money. The other way is first to add the two *lists*, obtaining a single list which gives his purchases over the two weeks:

> 3 quarts skim milk
> 3 candles
> 1 box napkins
> 60 hard rolls
> 7 frozen dinners

and then to multiply the cost per item by the numbers in this list. The total will be the amount he spends in the two weeks.

In many applications of mathematics today, these operations of adding lists together and multiplying one list by another are very common. From the observation that such operations are performed so often was born the idea of *matrix arithmetic*. The basic idea of matrix arithmetic is very simple—we treat the lists themselves as numbers, so that we imagine ourselves to be adding and multiplying these "numbers" together. This is so simple and so slight a change of viewpoint that a person at first is doubtful that it could lead to anything. However, it has proved to be one of the most fruitful ideas in all of mathematics.

The same kinds of addition and multiplication that can be performed on single lists can also be performed on rectangular arrays of numbers, as we will see below.

DEFINITION. An $m \times n$ *matrix* is any rectangular array of numbers having m rows and n columns. The number in the ith row from the top and in the jth column from the left is called the (i,j)-*entry* of the matrix.

For example, the array

$$\begin{pmatrix} 1 & 2 & 3 \\ 4 & 5 & 6 \end{pmatrix}$$

is a 2×3 matrix. Its $(2, 1)$-entry is the number 4. (Giant parentheses are usually employed to set off a matrix, but they are not absolutely necessary. Their main purpose is to indicate where one matrix ends and the next one begins.)

Addition of matrices is defined as follows.

DEFINITION. If A and B are both $m \times n$ matrices, their sum $A + B$ is that $m \times n$ matrix whose (i,j)-entry is the sum of the (i,j)-entries of A and B. If A and B are matrices of different sizes, the sum $A + B$ is not defined.

For example,

$$\begin{pmatrix} 1 & 2 & 3 \\ 4 & 5 & 6 \end{pmatrix} + \begin{pmatrix} 0 & 5 & -1 \\ 3 & 4 & 2 \end{pmatrix} = \begin{pmatrix} 1 & 7 & 2 \\ 7 & 9 & 8 \end{pmatrix}.$$

$$\begin{pmatrix} 1 & 2 & 3 \\ 4 & 5 & 6 \end{pmatrix} + \begin{pmatrix} 0 & 5 \\ 3 & 4 \end{pmatrix} \quad \text{is not defined.}$$

Multiplication of matrices is defined as follows.

DEFINITION. Let A be an $m \times k$ matrix and let B be a $k \times n$ matrix. The product AB is that $m \times n$ matrix that has for its (i,j)-entry the sum of the products of the ith row entries of A with the jth column

1. INTRODUCTION TO MATRICES

entries of B. If the number of columns of A is not equal to the number of rows of B, the product AB is not defined.

For example, the $(2, 1)$-entry of the product in Figure 2 is

$$4 \cdot (-1) + 5 \cdot 0 + 6 \cdot \tfrac{1}{3} = -2.$$

$$\begin{pmatrix} 1 & 2 & 3 \\ 4 & 5 & 6 \end{pmatrix} \begin{pmatrix} -1 & 3 & 2 \\ 0 & 1 & \tfrac{1}{2} \\ \tfrac{1}{3} & 2 & 0 \end{pmatrix} = \begin{pmatrix} 0 & 11 & 3 \\ -2 & 29 & 10\tfrac{1}{2} \end{pmatrix}$$

FIGURE 2. Illustration of matrix multiplication. To get the entry in the *second* row, *first* column of the product, we list-multiply the *second* row of the first matrix times the *first* column of the second matrix.

We obtain each entry of the product AB by multiplying two lists together, then summing, just as we did with the prices and the shopping list. Matrix multiplication is a generalization of list multiplication to rectangular arrays of numbers.

Of the two matrix operations, addition and multiplication, multiplication is by far the more interesting and useful. In most applications of matrices, the operation of multiplication that we have defined corresponds exactly to the physical or biological or sociological (or whatever) process we would most like to study. This comment, vague as it is now, should be kept in mind throughout this chapter. All three uses of matrices we will study involve multiplication very heavily, but in only one of them will we use addition.

The operation of addition is always of secondary consideration. What is most important about it is that it behaves "nicely" when mixed in with matrix multiplication. For example, it turns out that the operations of matrix multiplication and addition satisfy the following equations, known as the *distributive laws*:

$$A(B + C) = AB + AC, \qquad (1)$$
$$(A + B)C = AC + BC, \qquad (2)$$

whenever the products are defined. A special case of the first distributive law was our observation that we could find how much the shopper spent in the two weeks by (a) adding his two shopping lists B and C together, then taking the product $A(B + C)$, or (b) finding how much he spent each week, AB and AC, and then adding the expenditures together, $AB + AC$.

Three other laws which the operations of matrix addition and multiplication obey are

$$A + (B + C) = (A + B) + C, \qquad (3)$$
$$A(BC) = (AB)C, \qquad (4)$$
$$A + B = B + A. \qquad (5)$$

Equations (3) and (4) are called *associative laws*. When three matrices are to be added or multiplied, it does not matter whether the operation on the right is performed first, as we are directed to do when the parentheses are around B and C, or whether the operation on the left is performed first, as we are directed to do when the parentheses are around A and B. This amounts to telling us that we are allowed to *forget* parentheses in any long string of additions and in any long string of multiplications. It can be very handy at times to be able to get rid of parentheses which otherwise would clutter up formulas.

Equation (5) is called a *commutative law*.

All five of the equations we have written down are fairly easy to prove. The ingredients needed for the proofs are the definitions of matrix addition and multiplication, plus the ordinary laws (or axioms) of arithmetic which hold for the numbers appearing as the matrix entries. To be sure, it is difficult keeping track of all the entries, especially in proving (4), but that is the only difficulty which arises. Because the proofs are easy but "sticky," we will not give them here.

Some people get very annoyed at such proofs because they regard them as a lot of work for no benefit. They feel that Equations (1) through (5) are so simple that everyone should be willing to accept them without proof. However, that can be dangerous. Let us consider one more possibility of a law: Why did we not write down a commutative law for matrix multiplication, if for no other reason than to round out our list? The reason we did not is that matrix multiplication *does not obey* a commutative law. Compute the two products

$$\begin{pmatrix} 1 & 2 \\ 3 & 4 \end{pmatrix} \begin{pmatrix} 1 & 3 \\ 2 & 4 \end{pmatrix}$$

and

$$\begin{pmatrix} 1 & 3 \\ 2 & 4 \end{pmatrix} \begin{pmatrix} 1 & 2 \\ 3 & 4 \end{pmatrix}$$

and you will see that they are not the same. In fact, they differ in all four entries.

Since a very familiar property of multiplication of numbers does not carry over to multiplication of matrices, it really is necessary for a mathematician sometime in his career to determine which laws of ordinary arithmetic carry over to matrix arithmetic and which do not. If we were to do that here, however, we would have a chapter full of proofs with no room for applications of matrices. To compromise, we will give no proofs, but we will limit ourselves to using only Equations (1) through (5). That way, if you should ever wish to put our work in this chapter on a firm foundation, all you will need to do is prove those five equations.

EXERCISES

1. Calculate, in both the ways described in the text, the amount of money spent by the shopper.

2. Compute the products

$$\begin{pmatrix} 1 & 2 \\ 3 & 4 \end{pmatrix} \begin{pmatrix} 1 & 3 \\ 2 & 4 \end{pmatrix} \quad \text{and} \quad \begin{pmatrix} 1 & 3 \\ 2 & 4 \end{pmatrix} \begin{pmatrix} 1 & 2 \\ 3 & 4 \end{pmatrix}$$

so that you can see they are not the same.

3. Find two 3×3 matrices A and B such that AB is not equal to BA.

4. Let

$$A = \begin{pmatrix} 1 & 2 \\ 3 & 4 \end{pmatrix} \quad B = \begin{pmatrix} 1 & 2 \\ 1 & 5 \end{pmatrix} \quad C = \begin{pmatrix} 0 & 1 \\ 10 & -2 \end{pmatrix}.$$

Calculate the two products $A(BC)$ and $(AB)C$, both to build your skill in multiplying matrices and to verify that they are the same.

5. For the matrices A, B, and C of Exercise 4, compute

$$A(B+C) \quad AB + AC \quad (A+B)C \quad AC + BC.$$

6. Compute the products $A(BC)$ and $(AB)C$ for three 3×3 matrices A, B, and C of your own choosing.

7. Give an example of two 2×2 matrices A and B such that

$$AB = \begin{pmatrix} 0 & 0 \\ 0 & 0 \end{pmatrix}$$

but

$$A \neq \begin{pmatrix} 0 & 0 \\ 0 & 0 \end{pmatrix} \quad B \neq \begin{pmatrix} 0 & 0 \\ 0 & 0 \end{pmatrix}.$$

8. If $AB = AC$, can you then conclude that $B = C$? Prove your answer. [*Hint:* You can make use of Exercise 7.]

9. Give four different examples of a 2×2 matrix A such that

$$AA = A.$$

How many numbers x satisfy the equation

$$x^2 = x?$$

10. Let

$$A = \begin{pmatrix} .4 & .6 \\ .6 & .4 \end{pmatrix}.$$

Compute the products AA, AAA, $AAAA$, and $AAAAA$. If you see something happening that you think is worth mentioning, tell what that something is.

11. Repeat Exercise 10 with a 2×2 matrix of your own choosing. Do you see the same phenomenon happening this time?

2. The Adjacency Matrix of a Graph

Our first application of matrices will be to the study of graphs.

DEFINITION. A *graph* is any finite set of points (called *vertices*) together with certain line segments or arcs (called *edges*) joining these points. Certain of the edges may be assigned directions, to allow them to be traced only in the direction indicated.

You will recall we have already spent considerable time studying graphs in Chapter 4. In this chapter we will study them from a different point of view. We have repeated the definition of a graph here, slightly generalized, to make this chapter self-contained and also to give ourselves the opportunity of studying a more general kind of graph. Matrices lend themselves naturally to the study of this more general kind of graph.

It will prove helpful for us to think of a graph as a kind of road map. The vertices represent places to travel to, and the edges represent the different routes along which a person can travel. Just as within cities some streets may be one-way, so too in a graph certain edges may have directions assigned to them, to allow travel along them only in the direction indicated by the arrow. An example of a

2. THE ADJACENCY MATRIX OF A GRAPH

4-vertex graph, complete with several directed edges, is shown in Figure 3.

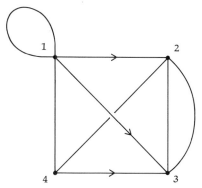

FIGURE 3. A 4-vertex graph, complete with "one-way streets," a "scenic loop," an "over-underpass," and two "highways" joining the same pair of vertices.

Notice from the example that it is possible for: (1) an edge to have both its ends at the same vertex, (2) 2 edges to cross each other without there necessarily being a way to get from one edge to the other, and (3) a pair of vertices to be joined by several edges. The corresponding phenomena in roads are: scenic loops, over-underpasses, and several different highways joining the same pair of cities.

Our next step is to define a matrix associated with a graph. Notice that we have already assigned numbers to the vertices of our graph in Figure 3. Numbered vertices are needed before the matrix can be defined.

DEFINITION. Let G be any graph whose n vertices have been numbered $1, 2, 3, \cdots, n$ in some order. The *adjacency matrix* A of G is the $n \times n$ matrix whose (i, j)-entry is the number of edges along which one can travel from vertex i to vertex j.

Using the numbers already assigned to our graph in Figure 3, we see that our graph has for its adjacency matrix the 4×4 matrix

$$A = \begin{pmatrix} 2 & 1 & 1 & 1 \\ 0 & 0 & 2 & 1 \\ 0 & 2 & 0 & 0 \\ 1 & 1 & 1 & 0 \end{pmatrix}.$$

You should check this matrix carefully against Figure 3 to see for yourself that the entries do give the number of edges leading from vertex to vertex. The "scenic loop" from vertex 1 to vertex 1 is counted as 2 edges, because it can be traveled in both clockwise and counterclockwise directions.

Let us consider a path leading through the graph along various edges. If the path travels along d edges all told (counting repetitions if there are any), we will say that the path has *length d*. For example, the path traced through the graph in Figure 4 is a path of length 3.

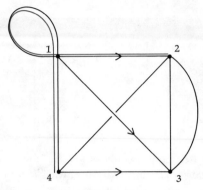

FIGURE 4. An example of a length 3 path through a graph. One can travel along the path from vertex 4 to vertex 2, but it is impossible to travel back along the path from vertex 2 to vertex 4, due to the directed edge.

We can think of the adjacency matrix as giving the number of length 1 paths from each vertex i to each vertex j. In certain applications of graphs, it is necessary to know the number of length 2 paths, the number of length 3 paths, and so on from each vertex i to each vertex j. Also, one frequently needs to know what the greatest path length is which will ever be needed in order to travel from any vertex i to any vertex j. The answers to both these problems can be found simply by performing some arithmetic on the adjacency matrix of the graph.

For example, let us consider how we might count the number of length 2 paths from vertex 1 to vertex 2 in the graph of Figure 3. To make the count, we first find the number of edges from vertex 1 to each of the vertices 1, 2, 3, and 4. These numbers are the numbers in the first row

$$2 \quad 1 \quad 1 \quad 1$$

of the adjacency matrix A. Then we find the number of edges from the vertices 1, 2, 3, and 4 to the vertex 2. These numbers are the numbers in the second column

$$\begin{matrix} 1 \\ 0 \\ 2 \\ 1 \end{matrix}$$

of the adjacency matrix A.

2. THE ADJACENCY MATRIX OF A GRAPH

To find the number of paths of length 2 from vertex 1 to vertex 2, we multiply the corresponding numbers and then add the four products:

Number of length 2 paths passing through:

$$
\begin{array}{rl}
\text{vertex 1:} & 2 \times 1 = 2 \\
\text{vertex 2:} & 1 \times 0 = 0 \\
\text{vertex 3:} & 1 \times 2 = 2 \\
\text{vertex 4:} & 1 \times 1 = \underline{1} \\
\text{Total number of paths:} & 5
\end{array}
$$

But this sum of products is by definition the $(1, 2)$-entry of the product AA. By a similar argument, the number of length 2 paths from vertex i to vertex j is always equal to the (i,j)-entry of the product AA. Now how do you suppose we might find the number of paths of length 3 from vertex i to vertex j? If we consider each path of length 3 to consist of two pieces, a path of length 1 followed by a path of length 2, we come to the conclusion that the number of length 3 paths from vertex i to vertex j is the (i,j)-entry of the product $A(AA)$.

Let us agree to denote the product of k A's by A^k. That is,

$$A^2 = AA,$$
$$A^3 = AAA,$$
$$A^4 = AAAA,$$

and so on. There is no danger of ambiguity in defining A^k, because the associative law (4) tells us we can forget parentheses. In general, the number of paths of length k from vertex i to vertex j will always be the (i,j)-entry of the matrix A^k. This is the answer to the first problem we mentioned.

The second problem was to find the greatest path length which will ever be needed in order to travel from one vertex to another. How do we find this greatest path length? Well, the (i,j)-entry of the sum $A + A^2$ counts the number of paths of length either 1 or 2 from vertex i to vertex j. The (i,j)-entry of the sum $A + A^2 + A^3$ counts the number of paths of length 1, 2, or 3 from vertex i to vertex j. And so on. As soon as we find a k such that all the (i,j)-entries, i unequal to j, of

$$A + A^2 + A^3 + \cdots + A^k$$

are nonzero, we know that every vertex i is joined to every vertex j by a path of length k or less. (The (i,i)-entries of a matrix are called its *diagonal* entries. The diagonal entries need not be greater than 0, because to get from vertex i to vertex i, it is not necessary to do any traveling; just stay where you are.)

For example, if we let A be the adjacency matrix of the graph in Figure 3, we find that

$$A^2 = \begin{pmatrix} 5 & 5 & 5 & 3 \\ 1 & 5 & 1 & 0 \\ 0 & 0 & 4 & 2 \\ 2 & 3 & 3 & 2 \end{pmatrix}$$

and therefore

$$A + A^2 = \begin{pmatrix} 7 & 6 & 6 & 4 \\ 1 & 5 & 3 & 1 \\ 0 & 2 & 4 & 2 \\ 3 & 4 & 4 & 2 \end{pmatrix}.$$

Since the $(3, 1)$-entry of $A + A^2$ is 0, we cannot travel from vertex 3 to vertex 1 by a path of length 2 or less. Continuing, then, we calculate

$$A^3 = \begin{pmatrix} 13 & 18 & 18 & 10 \\ 2 & 3 & 11 & 6 \\ 2 & 10 & 2 & 0 \\ 6 & 10 & 10 & 5 \end{pmatrix}$$

and therefore

$$A + A^2 + A^3 = \begin{pmatrix} 20 & 24 & 24 & 14 \\ 3 & 8 & 14 & 7 \\ 2 & 12 & 6 & 2 \\ 9 & 14 & 14 & 7 \end{pmatrix}.$$

This time all entries are different from 0, so we know that 3 is the longest path length we will ever need in order to travel from one vertex to another.

In this section we have seen only the beginning of a long and fruitful association of graphs and matrices. The further one goes into the subject, the more practical it becomes. For instance, as a next step one can assign numbers to the various edges to represent miles or traveling times. With this addition the graph becomes a very practical road map. The problem of finding the shortest route in time between two cities is an important one in the shipping industry. Another important problem, called the traveling salesman problem, is to find the shortest path through a graph which passes through *all* the vertices. The PERT chart, which we studied in Chapter 4, is another example of a graph with numbers assigned to its edges. In analyzing a

2. THE ADJACENCY MATRIX OF A GRAPH

PERT chart, the problem is to find the *longest* path (the critical path) from the end-of-project vertex back through the chart.

EXERCISES

1. In Figure 3, find a path of length 3 which leads from vertex 3 to vertex 1.
2. For Figure 3, the $(4, 2)$-entry of A^2 is equal to 3. Therefore there must be 3 paths of length 2 from vertex 4 to vertex 2. Find all 3 of them.
3. For each pair of vertices i, j in Figure 3 other than the pair 3, 1 find a path of length 1 or 2 from vertex i to vertex j. (On account of the directed edges, you may need to use a different path to go from j to i than you use to go from i to j.)
4. Write down the adjacency matrices for the following graphs.

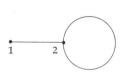 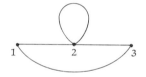

5. The graphs below were obtained from those in Exercise 4 by making some of the edges directed. Write down the adjacency matrices for these new graphs.

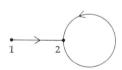 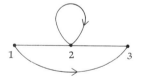

6. Write down the adjacency matrix for Figure 6 of Chapter 4. (Label the vertices 1, 2, 3, and 4, respectively, instead of A, B, C, and D.)
7. Use the matrix you wrote down in Exercise 6 to compute how many different paths of length 4 go from vertex 1 to vertex 3.
8. (For those who have read Chapter 4.) Sum each of the columns of the adjacency matrix in Exercise 6. What do the column sums mean in terms of the vertices and edges of the graph? Restate Theorem 1 of Chapter 4 in terms of adjacency matrices.

164   CHAP. 8 MATRICES

9. Let
$$A = \begin{pmatrix} 0 & 0 & 0 \\ 1 & 0 & 2 \\ 0 & 3 & 0 \end{pmatrix}$$
be the adjacency matrix of a graph which has 2 undirected edges between vertices 2 and 3. Draw the graph for which A is the adjacency matrix.

10. Let A be as in Exercise 9. Calculate $A + A^2$, $A + A^2 + A^3$, and $A + A^2 + A^3 + A^4$. Is there a natural number k such that all the (i,j)-entries, i unequal to j, of $A + A^2 + \cdots + A^k$ are positive? Explain why or why not from the graph you drew in Exercise 9.

11. Give an example of a graph for which no power of the adjacency matrix A has all off-diagonal entries positive, but for which all off-diagonal entries of $A + A^2$ are positive. [*Hint:* A graph with 3 vertices will do the trick. Try a few 3-vertex graphs until you find one that works.]

12. Let A be the adjacency matrix of a graph with n vertices. Show that if the sum $A + A^2 + \cdots + A^{n-1}$ has an off-diagonal entry equal to 0, then $A + A^2 + \cdots + A^{k-1}$ will have that same entry equal to 0, no matter how large k is.

13. Show that if A is the adjacency matrix of a graph with n vertices, then any vertex can be reached from any vertex if and only if each off-diagonal entry of $A + A^2 + \cdots + A^{n-1}$ is positive. [*Hint:* Use Exercise 12.]

3. Predicting Populations

In this section we will use matrices to develop a model for the growth (or possibly decline) of a population. It will become clear as we go along that this particular model is only a crude, first approximation to reality. The growth predictions it makes may turn out to be very rough indeed. We might think of this model as a baby's first tottering step—its greatest value being that it promises much better things to come.

We will begin by giving some fictitious population figures for the women of a certain country. The figures actually are based on census and vital statistics for the United States, but they have been rounded off in order to make the computations easier to follow. The population figures we will use are given in Table 1.

3. PREDICTING POPULATIONS

TABLE 1. Fictitious population statistics for the women of a certain country, of ages under 45 years.

Age Group	Number in Group in 1940	Number in Group in 1955	Girls Born to Each 1940 Age Group Surviving until 1955
0 to 14 years	14.5 million	16.4 million	4.6 million
15 to 29 years	15.3 million	14.3 million	10.4 million
30 to 44 years	11.3 million	14.8 million	1.4 million

What we are interested in doing is finding a matrix equation which will convert the column of 1940 population figures

$$\begin{pmatrix} 14.5 \\ 15.3 \\ 11.3 \end{pmatrix} \text{ million}$$

to the column of 1955 population figures

$$\begin{pmatrix} 16.4 \\ 14.3 \\ 14.8 \end{pmatrix} \text{ million.}$$

Let us first study how the 1955 figure of 14.8 million must have come about. This figure is the number of women aged 30 to 44 in the year 1955. In the year 1940, these women all belonged to the age group 15 to 29. If we neglect the effects of immigration and emigration, these women must then have been in the group of 15.3 million women of the country aged 15 to 29 in 1940. That is, of the 15.3 million women aged 15 to 29 in 1940, 14.8 million must have survived to the year 1955. The survival rate over 15 years for the group aged 15 to 29 in 1940 is thus

$$\frac{14.8}{15.3} = .967.$$

The same discussion applies to the 14.5 million women aged 0 to 14 in 1940. Of them, 14.3 million must have survived to the year 1955, to become the group of women aged 15 to 29 in 1955. The survival rate over 15 years for the group aged 0 to 14 in 1940 is thus

$$\frac{14.3}{14.5} = .986.$$

We are now left with the 16.4 million women aged 0 to 14 in 1955. They evidently were born sometime along the way from 1940, and so to account for them we need some birth information. This information appears in the fourth column of Table 1. Notice that the sum of births listed in that column is equal to 16.4 million. That is as it should be, since women of age 45 and older seldom bear children.

We have given the number of births for each of the three age groups in order to calculate the birth rate for each group over the 15-year period from 1940 to 1955. For the women of ages 0 to 14 in 1940, the birth rate is

$$\frac{4.6}{14.5} = .317.$$

(That is, on the average to each 1000 women aged 0 to 14 in 1940 there were 317 daughters born and surviving by the year 1955.) For the women of ages 15 to 29 in 1940, the birth rate is

$$\frac{10.4}{15.3} = .680.$$

And for the women of ages 30 to 44 in 1940, the birth rate is

$$\frac{1.4}{11.3} = .124.$$

We can now build a matrix from the survival and birth rates we have just calculated. By definition of these rates, we have the following three equations, which show how the population in 1955 comes from the population in 1940.

$$.317 \times 14.5 + .680 \times 15.3 + .124 \times 11.3 = 16.4$$
$$.986 \times 14.5 \qquad\qquad\qquad\qquad\qquad\qquad = 14.3$$
$$\qquad\qquad\quad .967 \times 15.3 \qquad\qquad\qquad\qquad = 14.8.$$

The left-hand side of the first equation is especially recognizable as an entry in a matrix product. Once we have noticed that, the pieces begin to fall into place. The left-hand sides of the next two equations are also entries in a matrix product, where one of the matrices in the product has a few 0's sprinkled around inside it. We combine the three equations into a single matrix equation:

$$\begin{pmatrix} .317 & .680 & .124 \\ .986 & 0 & 0 \\ 0 & .967 & 0 \end{pmatrix} \begin{pmatrix} 14.5 \\ 15.3 \\ 11.3 \end{pmatrix} = \begin{pmatrix} 16.4 \\ 14.3 \\ 14.8 \end{pmatrix}.$$

In what follows, we will denote the 3×3 matrix above by the letter T, to remind us of its role as a "transition" matrix.

So far we have not learned anything new about the population under study, because we have not yet gone beyond the years for which the information was given in Table 1. In order to go beyond,

3. PREDICTING POPULATIONS

we will need to make the rather daring assumption that birth rates and survival rates do not change from one 15-year period to the next. There is much to be said *against* making this assumption, but if we do not make it, we will get nowhere. So let us make it and see what happens. (We will discuss the shortcomings of our assumptions later.) We can then predict the number of women in each age group in 1970 by taking the product of T and the 1955 column matrix:

$$\begin{pmatrix} .317 & .680 & .124 \\ .986 & 0 & 0 \\ 0 & .967 & 0 \end{pmatrix} \begin{pmatrix} 16.4 \\ 14.3 \\ 14.8 \end{pmatrix} = \begin{pmatrix} 16.8 \\ 16.2 \\ 13.8 \end{pmatrix}.$$

Proceeding one step further, we can then predict the number of women in each age group in 1985 by:

$$\begin{pmatrix} .317 & .680 & .124 \\ .986 & 0 & 0 \\ 0 & .967 & 0 \end{pmatrix} \begin{pmatrix} 16.8 \\ 16.2 \\ 13.8 \end{pmatrix} = \begin{pmatrix} 18.1 \\ 16.6 \\ 15.7 \end{pmatrix}.$$

Another way to predict a future population is by first raising the transition matrix to an appropriate power T^k and then multiplying it times the 1940 population matrix

$$P = \begin{pmatrix} 14.5 \\ 15.3 \\ 11.3 \end{pmatrix}.$$

For instance, we can find the population matrix for the year 1970 by taking the product T^2P, and we can find the population matrix for the year 1985 by taking T^3P. The associative law of matrix multiplication guarantees that the only differences between the results found in this way and the results we found above will be the minor ones we ourselves introduce by rounding off.

It is an easy matter to extend the transition matrix to cover women belonging to the age groups 45 to 59, 60 to 74, and 75 to 89. The reason we did not do so here is that a 6 × 6 matrix is rather frightening—even to mathematicians. (There is, however, no particular reason to be frightened. Nowadays, most of the drudge work of matrix arithmetic is handled by computers, and a 6 × 6 and even a 60 × 60 matrix can easily be raised to powers by a moderate-sized computer.) By sticking to the population of women, we were justified in limiting our study to the ages under 45, because it is rare that women of age 45 and over bear children.

On the other hand, men of age 80 and beyond have been known to become fathers. We can guess, then, that predicting population for both sexes is a more complicated business. The model we have right now, however, is so rough that simply doubling all the population figures does not do much harm to its accuracy.

We have been admitting all along that our population growth model is not a very accurate one. There are far too many factors that can affect birth rates and survival rates for it to be useful for making predictions far into the future. It tells us how the population would continue to grow *if* the birth and survival rates did not change.

It is interesting to compare our model with a much earlier, oft-quoted model—that of Thomas Malthus (1766–1834). Malthus also assumed unchanging birth and survival rates. He concluded that population should increase "geometrically." Geometric increase means that the ratio of the population at the end of a period to the population at the beginning of that period should depend only on the length of the period, and not on the date the period began.

According to Malthus' model, the following population ratios should all be the same, since in each case the period length is 15 years.

$$\frac{1955 \text{ population}}{1940 \text{ population}} = \frac{16.4 + 14.3 + 14.8}{14.5 + 15.3 + 11.3} = \frac{45.5}{41.1},$$

$$\frac{1970 \text{ population}}{1955 \text{ population}} = \frac{16.8 + 16.2 + 13.8}{16.4 + 14.3 + 14.8} = \frac{46.8}{45.5},$$

$$\frac{1985 \text{ population}}{1970 \text{ population}} = \frac{18.1 + 16.6 + 15.7}{16.8 + 16.2 + 13.8} = \frac{50.4}{46.8}.$$

Performing the divisions, we find the ratios to be

$$1.107 \quad 1.029 \quad 1.077,$$

respectively, which are *not* very close to one another. The reason Malthus' model does not agree with ours is that he did not make use of the *different* birth and survival rates of the various age groups. His model is therefore less precise than ours. (By the same token, our model is less precise than one which would split the women under age 45 into 9 groups of ages 0 to 4, 5 to 9, 10 to 14, and so on.)

It is interesting, however, that after a number of periods have elapsed, the population ratios obtained from our model become quite stable, in line with Malthus' model. For example, with the aid of a computer I calculated T^{50}, T^{51}, T^{52}, and T^{53} and found that for the populations of women aged 0 to 44 predicted by our model for the years 2690, 2705, 2720, and 2735, the ratios were

$$\frac{2705 \text{ population}}{2690 \text{ population}} = \frac{717}{678} = 1.058,$$

3. PREDICTING POPULATIONS

$$\frac{2720 \text{ population}}{2705 \text{ population}} = \frac{758}{717} = 1.057,$$

$$\frac{2735 \text{ population}}{2720 \text{ population}} = \frac{801}{758} = 1.057,$$

which is just the sort of behavior Malthus expected.

Of course, neither the model of Malthus nor our model is a spectacularly good approximation to reality, because birth rates and survival rates do fluctuate for the reasons Malthus gave—overcrowding, disease, lack of food, and wars being the major ones. Ironically, the one humane means of limiting population that now seems to be working, contraception, was rejected by Malthus as an immoral solution.

EXERCISES

1. Use the model developed in the text to calculate the number of women aged 0 to 44 predicted for the years 2000, 2015, and 2030.

2. Calculate the predicted population ratios for the 15-year intervals 1985 to 2000, 2000 to 2015, and 2015 to 2030. Draw a graph to portray these ratios and the three ratios calculated in the text for the 15-year intervals between the years 1940 and 1985. Do the ratios seem to be settling down to a value near 1.057?

3. In the population figures for the year 1940 in Table 1, the number of women aged 0 to 14 is less than the number of women aged 15 to 29. This must mean that at least some of the years from 1925 to 1940 were years of low birth rates. Remembering that the figures of Table 1 are actually simplified United States population figures, can you think of any historical reason why birth rates should have been low then?

4. Having answered Exercise 3, consult a recent volume of the *Vital Statistics (Natality)* for the United States, published by the Department of Health, Education, and Welfare. Check the table listing birth rates for the individual years of this century to see if they support your answer to Exercise 3.

5. Assume some government agency is stockpiling dog-biscuits according to the following scheme. At present there are 1 billion fresh dog-biscuits in stock. At the end of each year, the number of fresh biscuits added to the stock will be 9/10 of the number in stock that year. Of the number of biscuits one year old, the

half lower in quality will be sold as surplus. Of the number of biscuits two years old, all will be sold. Write down the transition matrix for the stockpiling process. [*Hint:* Use a 2 × 2 matrix.]

6. Using the transition matrix from Exercise 5, find how many dog-biscuits will be in stock each year for the next five years. Find the ratio of the number each year to the number the previous year.

7. Repeat Exercises 5 and 6 with one modification: The number of fresh biscuits added to the stock will be 7/10 of the number in stock that year.

8. Let T and P be as in the text. Calculate $T^2 P$ and $T^3 P$. What differences are there between these two products and the two column matrices for 1970 and 1985 computed in the text?

9. By the associative law of matrix multiplication, T^{64} can be computed in only six multiplications:

$$T^{64} = (((((T^2)^2)^2)^2)^2)^2.$$

Using this technique, compute T^{64} for the transition matrix T of the text. Then compute $T^{65} = TT^{64}$. Use these two matrices to predict the number of women aged 0 to 44 in the years 2900 and 2915. What hs the ratio of these two numbers? How close is it to 1.057?

10. Certain plants, called *biennials*, germinate and grow leaves the first year, then flower and produce seeds the second year. After that they die. In a valley there are now 1,000,000 plants of one biennial species of wildflower—470,000 in their first year and 530,000 in their second. Under present conditions, each 1000 plants in their second year will, on the average, give rise to 1300 plants the following year, and of each 1000 plants in their first year, on the average 750 will survive through the second year. Write down the transition matrix for the population of wildflowers in the valley.

11. Continuing Exercise 10, predict the total number of wildflowers in the valley after 1, 2, 4, 8, 16, 17, and 18 years.

12. Repeat Exercise 11, except begin with 470,000 first-year plants and 230,000 second-year plants. (A natural catastrophe like a grass fire two winters before could have destroyed about half the standing plants while doing little damage to the seed about to germinate.)

13. From the results of Exercise 11, find the percent increase or decrease in the wildflower population from the 16th year to the

17th year. Find the percent increase or decrease from the 17th to the 18th year. Do the same for the results of Exercise 12. Does it look as if the growth in each case has settled down to follow Malthus' model?

4. Markov Chains

In this section we will study an application of matrices to probability theory. This application produces a model bearing many resemblances to the population growth model, the big difference being that the computations here will be done with probabilities instead of with numbers of people. The model is named in honor of the Russian mathematician A. A. Markov (1856–1922).

Suppose it is known that in a certain U.S. city having only the standard two parties, the Democrats have somewhat better luck getting members of their party elected mayor than do the Republicans. But suppose also that there is a "bandwagon" effect, so that each party stands a better chance of retaining power when incumbent than it does of taking power if it is not in office. For the sake of having a simple example, let us suppose that the various chances are the fractions given in Table 2.

TABLE 2. Probabilities of election of each party's candidate for mayor.

If the present mayor is Republican,	If the present mayor is Democrat,	
$\frac{2}{3}$	$\frac{1}{5}$	is the probability that the next mayor will be Republican
$\frac{1}{3}$	$\frac{4}{5}$	is the probability that the next mayor will be Democrat

For example, of all the times that an election is held when a Democrat is mayor, we expect that a Republican will be elected mayor $\frac{1}{5}$ of the times and that a Democrat will be elected $\frac{4}{5}$ of the times. The square array

$$M = \begin{pmatrix} \frac{2}{3} & \frac{1}{5} \\ \frac{1}{3} & \frac{4}{5} \end{pmatrix}$$

of numbers in Table 2 is called a *Markov matrix*.

Assume for the moment that we know the present mayor is Republican, and that an election is about to be held. The probabilities of election of a Republican and of a Democrat are given by the first column of the matrix M:

$$\begin{pmatrix} \frac{2}{3} \\ \frac{1}{3} \end{pmatrix}.$$

Now we can think of the state of affairs before the election as being represented by the column of probabilities

$$\begin{pmatrix} 1 \\ 0 \end{pmatrix}.$$

That is, since we *know* the present mayor is Republican, the probability that the Republicans presently hold power is 1, and the probability that the Democrats presently hold power is 0. Notice that in this case the post-election probabilities can be obtained by multiplying the pre-election probabilities by the Markov matrix:

$$\begin{pmatrix} \frac{2}{3} \\ \frac{1}{3} \end{pmatrix} = \begin{pmatrix} \frac{2}{3} & \frac{1}{5} \\ \frac{1}{3} & \frac{4}{5} \end{pmatrix} \begin{pmatrix} 1 \\ 0 \end{pmatrix}.$$

Similarly, in the case that we know the present mayor to be a Democrat, the post-election probabilities again can be obtained by multiplying the pre-election probabilities by the Markov matrix:

$$\begin{pmatrix} \frac{1}{5} \\ \frac{4}{5} \end{pmatrix} = \begin{pmatrix} \frac{2}{3} & \frac{1}{5} \\ \frac{1}{3} & \frac{4}{5} \end{pmatrix} \begin{pmatrix} 0 \\ 1 \end{pmatrix}.$$

These two special cases lead us to the very important idea: Suppose we do not know for certain who holds power before an election, but rather that the probabilities of each party being in power are

$$\begin{pmatrix} x \\ y \end{pmatrix} \quad \begin{matrix} \text{(Republican)} \\ \text{(Democrat)}. \end{matrix}$$

Is it still true that multiplication will give us the post-election probabilities?

$$\begin{matrix} \text{Post-election} \\ \text{probabilities} \end{matrix} \rightarrow \begin{pmatrix} x' \\ y' \end{pmatrix} = \begin{pmatrix} \frac{2}{3} & \frac{1}{5} \\ \frac{1}{3} & \frac{4}{5} \end{pmatrix} \begin{pmatrix} x \\ y \end{pmatrix}? \tag{6}$$

Let us see. The probability that the present mayor is Republican is x. If the present mayor *is* Republican, the probability that a Republican will be elected mayor is $\frac{2}{3}$. Therefore, the probability of the Republicans being in power both before *and* after the election is the product of these two probabilities: $\frac{2}{3}x$. The probability that the present mayor is a Democrat is y. If the present mayor *is* Democrat, the probability that a Republican will be elected mayor is $\frac{1}{5}$. Therefore, the probability of the Democrats being in power before the election *and* the Republicans being in power after the election is the product of these two probabilities: $\frac{1}{5}y$.

If the Republicans are elected to the mayoralty, then either they were reelected or they beat the incumbent Democrats. Therefore, the probability x' that the Republicans will be in power after the election is the sum of the probabilities of these two events:

$$x' = \tfrac{2}{3}x + \tfrac{1}{5}y. \tag{7}$$

Checking back to the matrix equation (6) which we suspect to be true, we find by taking the product on the right-hand side of (6) that Equation (7) is the upper half of the matrix equation (6).

In a similar manner, we can calculate that the probability y' of the Democrats being in power after the election is the sum

$$y' = \tfrac{1}{3}x + \tfrac{4}{5}y. \tag{8}$$

This equation is the lower half of the matrix equation (6), so we have now verified that (6) is correct.

(Admittedly, the last three paragraphs were heavy reading. It may be worthwhile to reread them, keeping an eye on Figure 5, which illustrates the discussion leading to Equation (7).)

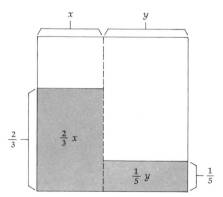

FIGURE 5. Pictorial representation of Equation (7). The shaded area on the left represents the probability the Republicans are in power and will be reelected; the shaded area on the right represents the probability the Democrats are in power and will lose to the Republicans. The entire shaded area, $\tfrac{2}{3}x + \tfrac{1}{5}y$, represents the probability the Republicans will win the next election.

Under what circumstances might we not know for certain which party is in power before an election? The most interesting case is when that election is some years in the future, so that there is at least one election between the present time and it. In such a case, we may already have used the Markov matrix to calculate probabilities for each party being in office before that election. According to (6), we can use the Markov matrix again, to calculate the probabilities for each party being in power after the election.

For example, if the present mayor is Republican, we know that the probabilities of each party holding power after the next election are

$$\begin{pmatrix} \frac{2}{3} & \frac{1}{5} \\ \frac{1}{3} & \frac{4}{5} \end{pmatrix} \begin{pmatrix} 1 \\ 0 \end{pmatrix} = \begin{pmatrix} \frac{2}{3} \\ \frac{1}{3} \end{pmatrix} = \begin{pmatrix} .667 \\ .333 \end{pmatrix}.$$

After the election following that, the probabilities of each party holding power are

$$\begin{pmatrix} \frac{2}{3} & \frac{1}{5} \\ \frac{1}{3} & \frac{4}{5} \end{pmatrix} \begin{pmatrix} \frac{2}{3} \\ \frac{1}{3} \end{pmatrix} = \begin{pmatrix} \frac{23}{45} \\ \frac{22}{45} \end{pmatrix} = \begin{pmatrix} .511 \\ .489 \end{pmatrix}.$$

After the third election, the probabilities of each party holding power are

$$\begin{pmatrix} \frac{2}{3} & \frac{1}{5} \\ \frac{1}{3} & \frac{4}{5} \end{pmatrix} \begin{pmatrix} \frac{23}{45} \\ \frac{22}{45} \end{pmatrix} = \begin{pmatrix} \frac{296}{675} \\ \frac{379}{675} \end{pmatrix} = \begin{pmatrix} .439 \\ .561 \end{pmatrix}.$$

(Notice, by the way, that in every column of probabilities

$$\begin{pmatrix} x \\ y \end{pmatrix}$$

so far, we have had

$$x + y = 1.$$

This will always happen, because at any time one or the other of the two parties must be in power.)

We can continue the above process of calculating probabilities for as many elections into the future as we like. Or, making use of the associative law for matrix multiplication, we can also find probabilities by raising the Markov matrix M to a power before multiplying it times the column of probabilities. For example, if we know that the Republicans are presently in power, the probabilities of each party being in power after the nth election can be found by multiplying:

$$\begin{pmatrix} \frac{2}{3} & \frac{1}{5} \\ \frac{1}{3} & \frac{4}{5} \end{pmatrix}^n \begin{pmatrix} 1 \\ 0 \end{pmatrix}.$$

Our election example is an example of a Markov chain in which there are two possible *states* at every stage—Republicans in power and Democrats in power. It is conceptually just as easy to make up Markov chain models for processes in which the same number s of states can appear at every stage, for any natural number s. The Markov matrix M is then an $s \times s$ matrix, and it is multiplied times columns of probabilities containing s entries. For instance, one might develop a Markov chain model for an election to an office for which there are more than two parties contending.

Or a military strategist might develop a Markov chain model to help himself understand troop levels in a certain foreign country. For example, 0 to 99,999 troops might be one state, 100,000 to 199,999 troops might be another state, and so on. He presumably would find his Markov matrix entries by estimating how much domestic and foreign pressures to expand or withdraw troops would be generated by each troop level state.

The Markov chain model can be used to calculate probabilities far into the future, but it must be used for that purpose with the same amount of caution we found appropriate in the case of the population growth model. Undoubtedly the probabilities within the usual Markov matrix are *not* constant throughout time; it is only reasonable to expect them to change somewhat, and any change in them represents a limitation in the accuracy of the model. As an example of the kinds of changes that can take place — and not in probabilities alone — recall that both our major parties are not much more than a century old, *and* that at their beginning, the Republican party was in general the more liberal of the two. The passing of one hundred years has so changed their roles that little more than the names remain the same.

Despite their shortcomings, Markov chains are one of those important steps in the right direction. As an important step, they have been investigated thoroughly. One striking property that has been proved to hold for a certain kind of Markov matrix M is a *stability property* of its powers. Whenever a Markov matrix M has the property that some power M^k of it has no entry equal to 0, it happens that higher and higher powers of M become closer and closer to a certain matrix L. The matrix L has the property that all of its columns are equal to each other.

Our example

$$M = \begin{pmatrix} \frac{2}{3} & \frac{1}{5} \\ \frac{1}{3} & \frac{4}{5} \end{pmatrix}$$

falls into this category, because every entry of M itself is different from 0. It turns out that as n increases, the powers

$$\begin{pmatrix} \frac{2}{3} & \frac{1}{5} \\ \frac{1}{3} & \frac{4}{5} \end{pmatrix}^n$$

become closer and closer to the matrix

$$L = \begin{pmatrix} \frac{3}{8} & \frac{3}{8} \\ \frac{5}{8} & \frac{5}{8} \end{pmatrix}.$$

(It is fairly easy to find L by solving a set of equations. The explanation will be given in Chapter 9.)

Notice that

$$\begin{pmatrix} \frac{3}{8} & \frac{3}{8} \\ \frac{5}{8} & \frac{5}{8} \end{pmatrix} \begin{pmatrix} 1 \\ 0 \end{pmatrix} = \begin{pmatrix} \frac{3}{8} \\ \frac{5}{8} \end{pmatrix}$$

and that

$$\begin{pmatrix} \frac{3}{8} & \frac{3}{8} \\ \frac{5}{8} & \frac{5}{8} \end{pmatrix} \begin{pmatrix} 0 \\ 1 \end{pmatrix} = \begin{pmatrix} \frac{3}{8} \\ \frac{5}{8} \end{pmatrix}.$$

Because high powers of M are close to the matrix L, we have that when n is large

$$\begin{pmatrix} \frac{2}{3} & \frac{1}{5} \\ \frac{1}{3} & \frac{4}{5} \end{pmatrix}^n \begin{pmatrix} 1 \\ 0 \end{pmatrix} \text{ approximately equals } \begin{pmatrix} \frac{3}{8} \\ \frac{5}{8} \end{pmatrix}$$

and also

$$\begin{pmatrix} \frac{2}{3} & \frac{1}{5} \\ \frac{1}{3} & \frac{4}{5} \end{pmatrix}^n \begin{pmatrix} 0 \\ 1 \end{pmatrix} \text{ approximately equals } \begin{pmatrix} \frac{3}{8} \\ \frac{5}{8} \end{pmatrix}.$$

That is, regardless of who holds office at present, Republican or Democrat, after many elections have been held, the outcome of the nth election will be almost completely independent of who is in power now. The Republicans will stand approximately $\frac{3}{8}$ chance of being elected, and the Democrats will stand approximately $\frac{5}{8}$ chance of being elected. We say that the Markov chain becomes *stable*.

We have now come to the end of our examples of the uses of matrices. The next chapter is an exposition of the first important tool a person must use in studying matrix theory. In this chapter, you have seen matrices used in three ways as models of certain aspects of reality. Very frequently the role of applied mathematics is to provide such models. Also very frequently such models fit reality only roughly, the reason being that they are simple enough to perform calculations on and therefore too simple to take into account the complexity of the real world. No one is particularly ashamed to admit the roughness of fit of his models. Study of a rough model can give understanding of a portion of reality—better than no understanding at all.

The further one goes into developing models, the better the models become. The reason is that the more complicated a model is, the more accurately it can be made to mirror reality. Our models are the very simplest, due to our being beginners. Many models studied nowadays are so complex that a human being alone cannot make predictions from them; the number of computations required is too great for him to make them without aid. Fortunately, the kinds of computations needed are usually the very repetitive kinds that digital computers excel at, so in the last two decades the study of mathematical models has been pushed forward rapidly.

EXERCISES

1. Round off the entries in the Markov matrix in the text to 3 decimal places:

$$M = \begin{pmatrix} .667 & .200 \\ .333 & .800 \end{pmatrix}.$$

   Then calculate M^2, M^4, M^8, and M^{16}, rounding to 3 decimal places each time. [*Hint:* Only 4 multiplications are required all told.] How close is M^{16} to the matrix L?

2. Figure 5 in the text illustrates the reasoning which leads to Equation (7). Draw a figure which illustrates the reasoning which leads to Equation (8).

3. Continuing the example in the text, compute the probabilities of each party being in power after the fourth, fifth, and sixth elections. How close are these probabilities to the column matrix

$$\begin{pmatrix} \frac{3}{8} \\ \frac{5}{8} \end{pmatrix}?$$

4. Using the Markov matrix in the text, and assuming that the Democrats presently hold power, compute the probabilities of each party being in power after each of the next six elections. Compare your results with the probabilities computed in the text and in Exercise 3 for the case where the Republicans presently hold power.

5. Suppose that every month in a certain "trouble spot" in the world there is the possibility of an "incident" occurring. Suppose that if an incident takes place during the present month, the probability of an incident occurring in the next month is $\frac{3}{4}$, while if an incident does not take place this month, the probability of an incident next month is $\frac{1}{2}$. Write down the Markov matrix for this process. Suppose an incident takes place this month. What is the probability there will be an incident three months from now?

6. In Exercise 5, suppose that an incident does not take place this month. What is the probability that an incident will occur three months from now?

7. Suppose two people play a game in which at each stage the loser pays the winner a dollar, and that at each stage each player has probability $\frac{1}{2}$ of being the winner. When someone loses all his money, the game ends. This game can be described by using an $(n + 1) \times (n + 1)$ Markov matrix, where n is the total number of dollars held by the players. Suppose $n = 3$. Write down the 4×4 Markov matrix M for the game. If one player begins with 1 dollar and the other with two dollars, what is the probability that after 4 stages the player starting with 2 dollars has won all the money? What is the probability that after 4 stages the player starting with 1 dollar has won all the money? What is the probability that neither player has won all the money after 4 stages?

8. Raise the Markov matrix of Exercise 7 to the 16th power. Although no power of M has all its entries greater than 0, it still happens that the powers M^k get close to some matrix L. From knowledge of M^{16}, make a guess as to what L is. What, then, are the probabilities that a person will eventually win all the money if he begins with 1 dollar? With 2 dollars? What is the probability the game will never end?

9. Write down the Markov matrix for the game of Exercise 7 in the case $n = 7$.

10. Since time immemorial, nations have engaged in competitions which bear remarkable resemblance to the game of Exercise 7. Pick a particular competition from history that you think resembles the game, and tell what features there are in common.

CHAPTER NINE

Row Reduction of Matrices

1. Introduction

The first great problem in matrix algebra is that of solving systems of *linear equations*. A *system of linear equations* with *unknowns* $x_1, x_2, \cdots, x_n$ is defined to be any series of equations of the form

$$a_{11}x_1 + a_{12}x_2 + \cdots + a_{1n}x_n + c_1 = 0$$
$$a_{21}x_1 + a_{22}x_2 + \cdots + a_{2n}x_n + c_2 = 0$$
$$\vdots \qquad \cdots$$
$$a_{m1}x_1 + a_{m2}x_2 + \cdots + a_{mn}x_n + c_m = 0,$$

where the a_{ij}'s and the c_i's are numbers like $2, -1, \frac{1}{2}, \sqrt{3}$, and so forth.

An apology is in order for such a massive dose of subscripts on the very first page of a chapter, but there simply is no other way to convey the notion of an arbitrary number m of equations, each with some number n of unknowns. As much as possible, we will avoid the use of subscripts in all that follows. An example of a system of 2 linear equations in 3 unknowns is

$$\frac{1}{2}x_1 + 3x_2 - 2x_3 + 2 = 0$$
$$1x_1 + 5x_2 - 4x_3 + 5 = 0. \qquad (\dagger)$$

No restriction is placed on the relative sizes of the number of equations m and the number of unknowns n. There may be more equations than unknowns, exactly the same number of equations as unknowns, or fewer equations than unknowns.

A *solution* of a system of linear equations is defined to be any series of n numbers which, when substituted *in order* for the unknowns $x_1, x_2, \cdots, x_n$ in each equation, produces the number 0 on the left-hand side of that equation. For example, the series of numbers

$$-2 \quad 1 \quad 2$$

is a solution to the system of equations (†) because

$$\tfrac{1}{2}(-2) + 3 \cdot 1 - 2 \cdot 2 + 2 = 0$$

and

$$1(-2) + 5 \cdot 1 - 4 \cdot 2 + 5 = 0.$$

Another solution to the system of Equations (†) is the series of numbers

$$-8 \quad 1 \quad \tfrac{1}{2}.$$

Notice that we do *not* consider the series of numbers

$$\tfrac{1}{2} \quad -8 \quad 1$$

to be a solution to (†), because when these numbers are substituted *in order* for x_1, x_2, and x_3, the left-hand sides of (†) do not become 0. That is, a rearrangement of a solution need not be a solution.

A given system of linear equations may have no solution at all, exactly one solution, or several (in fact, then, infinitely many) different solutions. The process of solving a system of linear equations is defined to be either finding all solutions to the system or else proving that the system has no solution. A method known as *row reduction* provides a foolproof, almost mechanical way to solve any system of linear equations.

It turns out, however, that row reduction's usefulness is not limited to solving equations. Once a person has mastered the technique, he can use it in many other ways in matrix algebra. It is the basic tool used in the study of the elementary side of matrix algebra; one can spend several months studying the subject using little else. The purpose of this chapter, then, is twofold. First of all, we will be studying row reduction as a means of equation solving. But second, we will be covering the basic technique used in studying the college mathematics subject now regarded as second only to calculus in its importance. Should you ever need to study matrix theory for use in your career, you will have received a headstart into the subject.

EXERCISES

1. Verify that the series of numbers
$$-8 \quad 1 \quad \tfrac{1}{2}$$
is a solution of the system (†).

2. Verify that the series of numbers
$$\tfrac{1}{2} \quad -8 \quad 1$$
is not a solution of the system (†).

2. Row Reduction by Example

The best way to learn row reduction is simply to go through a sufficiently complicated example, with actual numbers for the a_{ij}'s and c_i's. At the same time we work through the example, we will be giving reasons why the steps always work, so at the end we will have not only a completely worked example but also an illustrated proof that the row reduction technique always works. Let us take for our example system the three equations in four unknowns:

$$\left.\begin{aligned} 2x_1 + 4x_2 + x_3 - x_4 + 10 &= 0 \\ x_1 + 2x_2 + x_3 + 2x_4 &= 0 \\ -x_1 - 2x_2 + 4x_3 + x_4 + 3 &= 0 \end{aligned}\right\}. \tag{1}$$

(This example is sufficiently large that we can illustrate the full technique of row reduction on it. Because of its size, you may need a little patience to follow the process through to completion. The exercises at the end of the section will present no such problem. In them, you can build your skill gradually by working through them in order, from the simple ones to the more complicated ones.)

The first step in row reduction is to write down the matrix which is the array of coefficients of the system:

$$\begin{array}{rrrrr} 2 & 4 & 1 & -1 & 10 \\ 1 & 2 & 1 & 2 & 0 \\ -1 & -2 & 4 & 1 & 3. \end{array}$$

Notice that a few coefficients which were "understood" in the equations, like 1, −1, and 0, have to be supplied. The philosophy behind stripping away the unknowns is that one loses no information by doing so. The order in which the numbers appear is enough to tell us the unknowns to which they are attached. To recover the equations, all we need to do is put back the unknowns, some plus signs, and "= 0" on each line.

The second step is to add one column to this matrix, called the *check column*. The check column has nothing to do with the solution process but rather is used to guard against errors and locate them if they occur. It has proved to be a very dear friend of people who need to do complicated row reductions, because little mistakes are so easy to make and so hard to find. Each entry of the check column is defined to be the sum of all the entries of the corresponding row of the coefficient matrix. It appears as the right-hand column of the array below:

$$\begin{array}{rrrrrr} 2 & 4 & 1 & -1 & 10 & 16 \\ 1 & 2 & 1 & 2 & 0 & 6 \\ -1 & -2 & 4 & 1 & 3 & 5. \end{array}$$

We will call this matrix the *check-column matrix*.

We are now ready to begin the operations on the rows. If you were ever taught the addition-subtraction method of solving a system of equations, it will be helpful for you to keep in mind that the operations we will be performing on the rows of our matrix are very much the same as the manipulations used in the addition-subtraction method. If you know nothing about addition-subtraction, never fear; every step we perform will be explained carefully.

The overall plan is to convert the original system of equations into a simpler looking system without losing any solutions and without gaining any solutions in the process. It turns out that the conversion process can be carried so far that the new system of equations will be very, very simple—so simple that we will be able to write down the full set of solutions just by looking at the equations.

The steps to alter the equations are applied to the check-column matrix rather than to the equations themselves. The steps are known as *elementary row operations*.

An elementary row operation of the *first kind* consists in multiplying one row of the matrix by some number different from 0. The corresponding equations all stay the same, except for one, which has all its coefficients multiplied by the same nonzero number. It is fairly clear that this kind of operation does not cause any solutions to be lost, nor any solutions to be gained.

An elementary row operation of the *second kind* consists in switching two rows of the matrix. The corresponding equations are completely unchanged. The only change is in the order in which they are written down. In this case, too, the set of solutions remains unchanged.

An elementary row operation of the *third kind* consists in adding a multiple of one row to some other row. For definiteness, let us say that c times the ith row is being added to the jth row. All but one of the corresponding equations remain the same. The one that changes, the jth, is replaced by the sum of itself and c times the ith equation. Any solution of the original system of equations will therefore satisfy all the equations of the new system. On the other hand, the original matrix can be recovered from the new matrix by adding $-c$ times the ith row to the jth row, which is also an elementary row operation of the third kind. Therefore any solution of the new system of equations must be a solution of the original system as well. Therefore an elementary row operation of the third kind does not change the set of solutions to the system of equations.

We will now apply these three kinds of operations to change our check-column matrix into one which has a 1 in the upper left-hand corner and 0's everywhere else in the first column.

Switch the first and second rows:

$$\begin{array}{rrrrrr} 1 & 2 & 1 & 2 & 0 & 6 \\ 2 & 4 & 1 & -1 & 10 & 16 \\ -1 & -2 & 4 & 1 & 3 & 5. \end{array}$$

Add the first row to the third row:

$$\begin{array}{rrrrrr} 1 & 2 & 1 & 2 & 0 & 6 \\ 2 & 4 & 1 & -1 & 10 & 16 \\ 0 & 0 & 5 & 3 & 3 & 11. \end{array}$$

Add -2 times the first row to the second row:

$$\begin{array}{rrrrrr} 1 & 2 & 1 & 2 & 0 & 6 \\ 0 & 0 & -1 & -5 & 10 & 4 \\ 0 & 0 & 5 & 3 & 3 & 11. \end{array}$$

Let us remind ourselves at this point that, because we have used only the kinds of elementary row operations we defined earlier, the system of equations corresponding to this last matrix has exactly the same solutions as the original system. Notice, however, that the new system is simpler looking, because the second and third equations involve only the two unknowns x_3 and x_4.

We will now focus our attention on the smaller matrix obtained by deleting the first row and first column of the last matrix:

$$\begin{array}{rrrrr} 0 & -1 & -5 & 10 & 4 \\ 0 & 5 & 3 & 3 & 11. \end{array}$$

We locate the first column in this matrix which contains an entry other than 0. Having located the column, we apply elementary row operations to obtain a 1 in the top of the column in the smaller matrix and 0's everywhere else in the column in the full matrix.

Multiply the second row (of the full matrix) by -1:

$$\begin{array}{cccccc} 1 & 2 & 1 & 2 & 0 & 6 \\ 0 & 0 & 1 & 5 & -10 & -4 \\ 0 & 0 & 5 & 3 & 3 & 11. \end{array}$$

Add -1 times the second row to the first row:

$$\begin{array}{cccccc} 1 & 2 & 0 & -3 & 10 & 10 \\ 0 & 0 & 1 & 5 & -10 & -4 \\ 0 & 0 & 5 & 3 & 3 & 11. \end{array}$$

Add -5 times the second row to the third row:

$$\begin{array}{cccccc} 1 & 2 & 0 & -3 & 10 & 10 \\ 0 & 0 & 1 & 5 & -10 & -4 \\ 0 & 0 & 0 & -22 & 53 & 31. \end{array}$$

The strategy which begins to emerge from our example is this: Try to get each succeeding row headed by the number 1. Those number 1's will be called the *lead* 1's on account of their positions in the rows. Make all other entries in a column containing a lead 1 equal to 0.

There is now just one entry left which we would like to turn into a lead 1, so let us see to that. We focus our attention on the smaller matrix consisting of those entries both below and to the right of the last lead 1 we obtained:

$$\begin{array}{ccc} -22 & 53 & 31. \end{array}$$

Instead of the number -22, we wish to have the number 1, so we apply row operations to the full matrix to convert the -22 to a 1 and to make all other entries of that column equal to 0.

Multiply the third row (of the full matrix) by $-1/22$:

$$\begin{array}{cccccc} 1 & 2 & 0 & -3 & 10 & 10 \\ 0 & 0 & 1 & 5 & -10 & -4 \\ 0 & 0 & 0 & 1 & -53/22 & -31/22. \end{array}$$

Add 3 times the third row to the first row:

$$\begin{array}{cccccc} 1 & 2 & 0 & 0 & 61/22 & 127/22 \\ 0 & 0 & 1 & 5 & -10 & -4 \\ 0 & 0 & 0 & 1 & -53/22 & -31/22. \end{array}$$

Add -5 times the third row to the second row:

$$\begin{array}{cccccc} 1 & 2 & 0 & 0 & 61/22 & 127/22 \\ 0 & 0 & 1 & 0 & 45/22 & 67/22 \\ 0 & 0 & 0 & 1 & -53/22 & -31/22. \end{array}$$

With a larger matrix, this process of creating lead 1's and 0's is continued until finally there are no rows left in which to create a lead 1. When the process finally does come to an end, the matrix obtained is said to be in *row-reduced form*. A matrix in row-reduced form has the following properties: (1) Every row containing an entry different from 0 has the number 1 as its first nonzero entry; these 1's are what we are calling *lead* 1's. (2) Each column containing a lead 1 has all of its other entries equal to 0. (3) As one progresses down through the rows, the lead 1's appear farther and farther to the right.

It is easy to read off all the solutions to a system of equations whose matrix is in row-reduced form. For instance, the system of equations corresponding to our row-reduced matrix above is

$$\begin{aligned} x_1 + 2x_2 + 61/22 &= 0 \\ x_3 + 45/22 &= 0 \\ x_4 - 53/22 &= 0. \end{aligned}$$

The third equation tells us precisely what value x_4 must have: $53/22$, and the second equation tells us precisely what value x_3 must have: $-45/22$. The first equation tells us that we may let x_2 have any value c we wish, whereupon x_1 will be determined in terms of c:

$$\begin{aligned} x_1 &= -2x_2 - 61/22 \\ &= -2c - 61/22. \end{aligned}$$

Therefore, the full set of solutions to our system of equations (both this system and the system we began with) is

$$-2c - \frac{61}{22} \quad c \quad -\frac{45}{22} \quad \frac{53}{22}. \tag{2}$$

For instance, letting $c = -4$, we have a particular example of a solution:

$$\frac{115}{22} \quad -4 \quad -\frac{45}{22} \quad \frac{53}{22}. \tag{3}$$

Given any row-reduced matrix, we can get the complete set of solutions of the corresponding equations in the same way. Any unknown whose column does *not* contain a lead 1 can be set equal to any number we choose. Any unknown whose column *does* contain a lead 1 will appear in only one equation, with coefficient equal to 1. That equation determines its value in terms of the unknowns we were allowed to choose arbitrarily.

The process of reading off the solutions can lead to any of three situations: (1) no solutions, (2) exactly one solution, (3) infinitely many solutions. Let us see how each situation arises.

(1) It may have happened that after the row-reduction was completed, the second-to-last column — the one containing the constant terms c_i of the equations — contained a lead 1. The row containing that lead 1 would correspond to the equation

$$0x_1 + 0x_2 + \cdots + 0x_n + 1 = 0.$$

No series of values substituted for $x_1, x_2, \cdots, x_n$ can possibly satisfy this equation. Therefore the system of equations has no solutions.

(2) It may have happened that every column corresponding to an unknown contained a lead 1 and that the second-to-last column did not contain a lead 1. In that case, we have one equation for each unknown, telling us exactly what value that unknown must have.

(3) It may have happened that some column corresponding to an unknown x_i did not contain a lead 1 and that the second-to-last column did not contain a lead 1. That unknown can be assigned any value we please, so there are in this case an infinite number of solutions.

You will notice that so far we have made no use of the check column. The reason is that our example of row reduction contains no errors. However, when you try a complicated row reduction on your own, you will make errors. In fact, when I first did the row reduction in this section, I made two separate arithmetic mistakes — both silly little things. They would have been hard to find just by going over all the arithmetic, but with the check column I was able to locate them very quickly.

How is the column used? The principle behind its use is that all elementary row operations preserve the relation of the check column to the rest of the matrix. That is, each entry of the check column remains equal to the sum of all the other entries in its row. Every so often during a row reduction, you should sum each row except for the check column entry. If the sum is not exactly equal to the check column entry, you know you have made a mistake, and you should go back and correct it. It will have occurred somewhere between the last time you made the check and the current time.

The check should be performed often enough that you will not need to go back too far into the process to locate and correct your mistake. Some errors make it necessary to change all the row-reduction steps coming after them, so letting a check go for a long time may cause a lot of erasing and refiguring. One place at which it is convenient to make the check is immediately after the creation of a new lead 1. There should, of course, be a check at the very end.

Let us return to our row-reduction example and do a couple of the checks, just to get the feel of them. After we created the second lead 1:

$$1 + 2 + 1 + 2 + 0 = 6$$
$$0 + 0 + 1 + 5 - 10 = -4$$
$$0 + 0 + 5 + 3 + 3 = 11.$$

Thus everything checks at that stage. And at the very end:

$$1 + 2 + 0 + 0 + 61/22 = 127/22$$
$$0 + 0 + 1 + 0 + 45/22 = 67/22$$
$$0 + 0 + 0 + 1 - 53/22 = -31/22.$$

We should mention that the check column is not completely reliable. It is possible for two mistakes to cancel each other. For example, if one entry of a row has value 1 more than its true value and another has value one less than its true value, the check column will not detect the double mistake. (It is possible for one of the mistakes to be in the check column. It is also possible for a single mistake to occur in the check column; then the row reduction of the equations will be correct, but the check will indicate a mistake.) Experience has shown, however, that such perfectly matched mistakes are very rare. Therefore the check column is almost as good a device as a person could hope for.

EXERCISES

1. Verify that the series of numbers (3) satisfies the system of equations (1).

2. Choose three values of c other than -4 and verify for each of those values that the series of numbers (2) satisfies the system of equations (1).

3. Show that the series of numbers (2) satisfies the system of equations (1) for all values of c.

In the next five exercises the equations have been written in terms of the unknowns x, y, z in order to avoid the use of subscripts. This has no effect on the technique for solving them.

4. Use row reduction to solve the following systems of equations.
   (a) $x + y - 3 = 0$
   $\quad\; 2x - y\quad\; = 0.$
   (b) $x + y = 0$
   $\quad\; 3x + 3y = 0.$
   (c) $x + y - 1 = 0$
   $\quad\; 3x + 3y - 2 = 0.$

5. Use row reduction to solve the following systems of equations.
   (a) $x + 2y + 3z - 9 = 0$
   $\quad\; 3x + 6y + z - 11 = 0.$
   (b) $2x + y + 7z + 13 = 0$
   $\quad\; x + 2y + 5z + 14 = 0.$

6. Use row reduction to solve the system of equations:
   $$x + 2y + 3z - 4 = 0$$
   $$5x + 6y + 7z - 8 = 0$$
   $$9x + 10y + 11z - 12 = 0.$$

7. Use row reduction to solve the system of equations:
   $$5x = 8 - 7z - 6y$$
   $$2y = -x - 3z + 4$$
   $$11z = 12 - 10y - 9x.$$

8. Use row reduction to solve the system of equations:
   $$2x = 2y - 7z + 6$$
   $$x - 4y = 7$$
   $$6y = 3x + 3.5z + 3.5.$$

9. Make up a system of 3 linear equations in 3 unknowns by picking all the numbers x_{ij} and c_i to be different from 0, but otherwise random. Solve your system, taking special note of whether it has no solution, exactly one solution, or infinitely many solutions. You may be interested to know how many solutions your fellow classmates found for their systems.

10. Repeat Exercise 9 for the case of 4 equations in 3 unknowns. But also give an example of a system of 4 equations in 3 unknowns which has infinitely many solutions.

11. If in a system of linear equations all the c_i are equal to 0, can the system fail to have a solution?

3. Markov Chains Revisited

In the preceding chapter, we discussed a stability property of a certain kind of Markov matrix M. The stability property is that as n becomes larger and larger, the powers M^n become closer and closer to a matrix L. Let us agree to describe this behavior by saying that M^n *converges* to the matrix L. Previously, the best we could do toward finding the L for a given M was to raise M to powers 2, 3, 4, and so on until we could guess what M^n converged to. Now that we know how to solve systems of linear equations by row reduction, we will be able to develop a method for determining L exactly, *without* having to raise M to powers.

We will not, however, be able to prove that L exists, nor to prove that all its columns are the same. These conclusions require a lot of work to arrive at. We will simply assume them to be true and then make use of them to develop our method for finding L.

Because M^n converges to L, the high powers of M must all be approximately equal to each other, since they are all approximately equal to L. As a special case of this observation, we have

$$M^{n+1} \text{ approximately equals } M^n,$$

which is the same as

$$MM^n \text{ approximately equals } M^n. \qquad (4)$$

But also we have

$$M^n \text{ approximately equals } L. \qquad (5)$$

Putting the two approximations together, we have

$$ML \text{ approximately equals } L. \qquad (6)$$

Now, by making n sufficiently large, we can make the approximations (4) and (5) as close as we please. Therefore we can also make the approximation (6) as close as we please, just by making n sufficiently large. But n does not enter into (6) in any way, so the closeness of the approximation (6) is not changed by changing n. How, then, can (6) become as accurate as we like when neither side of it can be changed by varying n? The only way is if (6) is actually an equality. That is, it must be true that

$$ML = L. \qquad (7)$$

The matrix equation (7) furnishes us with a system of linear equations that we can use in finding the entries of L. In keeping with an earlier promise to avoid subscripts when possible, we will demonstrate how to solve for the entries of L by working through a reasonably-sized numerical example.

Let M be the 3×3 Markov matrix

$$M = \begin{pmatrix} \frac{1}{3} & \frac{1}{2} & 0 \\ \frac{1}{3} & \frac{1}{2} & \frac{1}{2} \\ \frac{1}{3} & 0 & \frac{1}{2} \end{pmatrix}.$$

You can quickly check that M^2 has no entry equal to 0, so the matrix L does exist. Let us denote L by

$$L = \begin{pmatrix} x & x & x \\ y & y & y \\ z & z & z \end{pmatrix}.$$

From the matrix equation

$$\begin{pmatrix} \frac{1}{3} & \frac{1}{2} & 0 \\ \frac{1}{3} & \frac{1}{2} & \frac{1}{2} \\ \frac{1}{3} & 0 & \frac{1}{2} \end{pmatrix} \begin{pmatrix} x & x & x \\ y & y & y \\ z & z & z \end{pmatrix} = \begin{pmatrix} x & x & x \\ y & y & y \\ z & z & z \end{pmatrix} \qquad (8)$$

we obtain the three equations

$$\begin{aligned} \tfrac{1}{3}x + \tfrac{1}{2}y + 0z &= x \\ \tfrac{1}{3}x + \tfrac{1}{2}y + \tfrac{1}{2}z &= y \\ \tfrac{1}{3}x + 0y + \tfrac{1}{2}z &= z. \end{aligned} \qquad (9)$$

These three equations, however, are not quite enough to determine x, y, and z. There is another condition on the unknowns which does not come from the matrix equation (7). That condition is

$$x + y + z = 1. \qquad (10)$$

To see where Equation (10) comes from, notice that each column of the matrix M^n, n any natural number, sums to 1. This fact can be seen without trying to multiply out M^n. The first column of M^n is equal to the product

$$M^n \begin{pmatrix} 1 \\ 0 \\ 0 \end{pmatrix}.$$

But the entries of this product are the three probabilities that after n stages the Markov process will be in each of the three possible states, given that it is now in the first state. Since the process must be in one of the three states at each stage, the sum of the three probabilities must be equal to 1. By a similar argument, the other columns of M^n also sum to 1. Since each column of M^n sums to 1, and since M^n converges to L, each column of L must also sum to 1.

We now combine Equations (9) and (10) into one system and try our hand at solving the system:

3. MARKOV CHAINS REVISITED

$$x + y + z - 1 = 0$$
$$-\tfrac{2}{3}x + \tfrac{1}{2}y + 0z + 0 = 0$$
$$\tfrac{1}{3}x - \tfrac{1}{2}y + \tfrac{1}{2}z + 0 = 0$$
$$\tfrac{1}{3}x + 0y - \tfrac{1}{2}z + 0 = 0.$$

The check column matrix of this system is

$$\begin{pmatrix} 1 & 1 & 1 & -1 & 2 \\ -\tfrac{2}{3} & \tfrac{1}{2} & 0 & 0 & -\tfrac{1}{6} \\ \tfrac{1}{3} & -\tfrac{1}{2} & \tfrac{1}{2} & 0 & \tfrac{1}{3} \\ \tfrac{1}{3} & 0 & -\tfrac{1}{2} & 0 & -\tfrac{1}{6}, \end{pmatrix} \qquad (11)$$

which row-reduces to

$$\begin{pmatrix} 1 & 0 & 0 & -\tfrac{1}{3} & \tfrac{2}{3} \\ 0 & 1 & 0 & -\tfrac{4}{9} & \tfrac{5}{9} \\ 0 & 0 & 1 & -\tfrac{2}{9} & \tfrac{7}{9} \\ 0 & 0 & 0 & 0 & 0. \end{pmatrix}$$

The system of equations corresponding to this row-reduced matrix is

$$x \qquad\qquad -\tfrac{1}{3} = 0$$
$$\qquad y \qquad -\tfrac{4}{9} = 0$$
$$\qquad\qquad z - \tfrac{2}{9} = 0.$$

Therefore the matrix L is

$$L = \begin{pmatrix} \tfrac{1}{3} & \tfrac{1}{3} & \tfrac{1}{3} \\ \tfrac{4}{9} & \tfrac{4}{9} & \tfrac{4}{9} \\ \tfrac{2}{9} & \tfrac{2}{9} & \tfrac{2}{9} \end{pmatrix}.$$

The procedure for finding L always works this same way, regardless of the size of M. From the equation

$$ML = L, \qquad (7)$$

you will be able to obtain as many different equations as M has rows. To these equations must always be added the equation which says that each column of L sums to the number 1. The combined system of equations can be solved by row reduction to find exactly one solution, which consists of the entries in each column of L.

EXERCISES

1. Row-reduce the check column matrix (11).

2. Raise the matrix M of the text to the powers M^2, M^4, M^8, M^{16}, and M^{32}. Compare these powers with the matrix L.

3. Give a proof, similar to the one in the text, that the second and third columns of the matrix M^n sum to 1.

4. Use the methods of this section to find L when
$$M = \begin{pmatrix} \frac{2}{3} & \frac{1}{5} \\ \frac{1}{3} & \frac{4}{5} \end{pmatrix}.$$
(Recall this is the Markov matrix studied in Section 4.)

5. Use the methods of this section to find L when
$$M = \begin{pmatrix} \frac{3}{4} & \frac{1}{2} \\ \frac{1}{4} & \frac{1}{2} \end{pmatrix}.$$

6. Raise the matrix M of Exercise 5 to the powers M^2, M^4, M^8, M^{16}, and M^{32}. Compare these powers with the matrix L.

7. Let
$$M = \begin{pmatrix} 0 & 1 & 0 \\ 0 & 0 & 1 \\ 1 & 0 & 0 \end{pmatrix}.$$
Does M^n converge to a matrix L? [*Hint:* Experiment, raising M to the powers M^2, M^3, M^4, and note what happens.] Such an M is called a *cyclic* Markov matrix.

8. For the matrix M in Exercise 7, try solving for L by the methods of this section, and then tell what problems (if any) arise in the solution process.

9. Let
$$M = \begin{pmatrix} 1 & \frac{1}{2} & 0 & 0 \\ 0 & 0 & \frac{1}{2} & 0 \\ 0 & \frac{1}{2} & 0 & 0 \\ 0 & 0 & \frac{1}{2} & 1 \end{pmatrix}.$$
Do you think M^n converges to a matrix L? [*Hint:* Experiment with powers of M. You should have no difficulty making an educated guess.] The matrix M, which you may have seen in a previous exercise, is an example of an *absorbing* Markov matrix.

10. For the matrix M in Exercise 9, try solving for L by the methods of this section, and then tell what problems arise in the solution process.

CHAPTER TEN

Introduction to Probability Theory

1. Introduction

Games of chance seem to be as old as mankind, but the mathematical theory of probability which studies them dates back only to the sixteenth century. Early writers on probability theory were Cardano, Fermat, and Pascal. Invention of calculus made possible further developments in the subject by a number of eighteenth century mathematicians.

Gradually, probability theory become less dependent upon its origins in card and dice games as it became developed sufficiently to be applied to more complex human endeavors. Then, as the amount of knowledge increased, investigation began to proceed along two separate lines. The more theoretical branch of the subject is still called probability theory, while the one which concerns itself with the more down-to-earth task of analyzing real-life data is known as mathematical statistics. Probability theory is concerned more with the study of mathematical models (as, for example, Markov chains), while mathematical statistics is more computational and more oriented toward obtaining answers to specific questions.

We are going to study first a bit of probability theory as groundwork and then proceed to study one of the main uses of mathematical statistics—hypothesis testing. In this topic, then, you will see a branch of applied mathematics which is oriented toward computing rather than studying models. The three hypothesis tests you will learn in Chapter 11 are among the most useful in statistics today. They are also quite modern in origin, the oldest having been devised less than 100 years ago and the youngest less than 30 years ago. As a byproduct of studying these tests, you will learn what a statistical proof is and how it differs from a mathematical proof.

The foundations of probability theory have been of great interest to mathematicians and philosophers since the very beginnings of the subject. At present there is a quite satisfactory set of definitions and axioms underlying the subject. However, to go into the foundations absorbs a lot of time. Instead, we will begin our study by assuming that you already have a vague notion of what is meant by the statement, "The probability of (blank) today is ___ percent." One hundred years ago this assumption would have been completely unwarranted, but in this age of mass media, nearly everyone has heard or read a weather report given in terms of probabilities and formed his own ideas of what the numbers called probabilities mean.

One popular way of thinking about probabilities is the so-called *frequency interpretation*. In the frequency interpretation, a statement like "The probability of a coin landing heads up is $\frac{1}{2}$" is replaced by a statement like "If a coin is flipped many times, we expect it to fall heads up $\frac{1}{2}$ of the time."

Probability computations are often easier to visualize when they are stated in frequency language. Suppose, for example, a person plays a game in which he must obtain a head by flipping a coin *and* roll a 6 on a die in order to win. If the coin and die are balanced, the probability of obtaining the head is $\frac{1}{2}$, and the probability of obtaining the 6 is $\frac{1}{6}$, so the probability of obtaining both the head and the 6 is

$$\tfrac{1}{2} \times \tfrac{1}{6} = \tfrac{1}{12}.$$

In terms of frequencies, we would rephrase the above statement as follows: If a person flips a balanced coin many times, he expects it to fall heads up $\frac{1}{2}$ of the time. If whenever he obtains a head he then rolls a balanced die, he expects to roll a 6 in $\frac{1}{6}$ of those times. Therefore, he expects to obtain both a head and a 6 in

$$\tfrac{1}{2} \times \tfrac{1}{6} = \tfrac{1}{12}$$

of the times he plays the game. That is, his probability of winning any one game is $\frac{1}{12}$.

All the frequency interpretation really does is throw the burden of understanding on the word *expect* instead of on *probability*. However, this shift of wording can help understanding, so there is good reason to use it whenever it is of help. You should feel free to rephrase

2. Flipping a Fair Coin

Probability theory can be a difficult, involved subject to learn. As much as possible, we will soften the difficulties by building our knowledge upon the simplest of all probability topics, the behavior of fair coins.

A *fair* coin is defined to be a coin that can land with either of two sides turned up, usually called *heads* and *tails*, such that the probability of its landing heads up is exactly $\frac{1}{2}$ and the probability of its landing tails up is exactly $\frac{1}{2}$. See Figure 1.

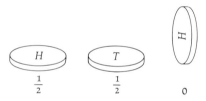

FIGURE 1. Probabilities for three positions of a fair coin.

There are two comments we should make about this definition. First, fairness is a restriction not only on the coin's structure, but also on the method of flipping the coin. If a coin is given no spin or very little spin, then it is more likely to fall with the same side up that was up when it was released. Second, this definition of fairness simply refers the idea of fairness one step further back, to the idea of probability. We leave the word probability without a formal definition, because an attempt to supply a definition would take us back into the foundations of probability theory.

A truly fair coin, of course, exists only in the mind. By the fact that the heads side of a coin is stamped differently from the tails side, there is bound to be a slight imbalance that will show up in a very large number of trials. Also, since every coin has some thickness, there is a faint chance it can land and remain standing on edge. (In Section 4, we will investigate the U.S. penny in regard to fairness. That study will be our first example of a hypothesis test.)

Suppose two fair coins are flipped. There are 4 distinct outcomes possible, because the first coin can turn up either heads or tails, and then in either of these 2 cases the second coin can turn up either heads or tails. We can use a branching or *tree* diagram to represent these 4 possibilities, as in Figure 2.

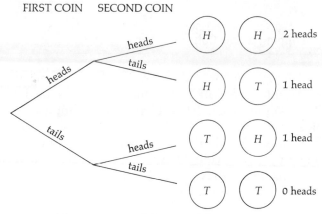

FIGURE 2. Possible outcomes of the act of flipping two coins.

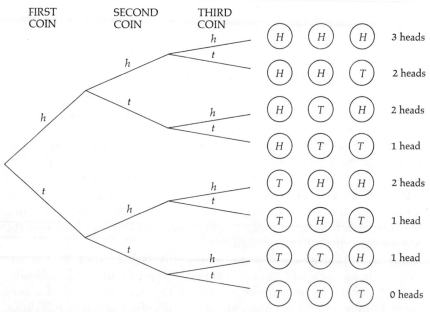

FIGURE 3. Possible outcomes of the act of flipping three coins.

Since each coin is assumed to be fair, each one of these 4 cases is equally likely to occur. That is, each case has probability $\frac{1}{4}$ of occurring. Now, of these 4 cases, there is 1 case with no heads showing, 2 cases with one head showing, and 1 case with two heads showing. Therefore, if we flip two fair coins, it is more likely that we will wind

2. FLIPPING A FAIR COIN   197

up with exactly one head showing (probability $\frac{1}{2}$) than we will wind up with no heads showing (probability $\frac{1}{4}$) or with two heads showing (probability $\frac{1}{4}$).

Let us carry on this same discussion for the case of 3 fair coins. There are 8 possible cases, all equally likely. These 8 cases are shown in the tree diagram of Figure 3. Of these 8 cases, there is 1 case with no heads showing, 3 cases with one head showing, 3 cases with two heads showing, and 1 case with three heads showing. Therefore, if we flip three fair coins, the probabilities of the outcomes no, one, two, and three heads are, respectively, $\frac{1}{8}$, $\frac{3}{8}$, $\frac{3}{8}$, and $\frac{1}{8}$.

Table 1 summarizes our discussion of flipping two and three fair coins. It also contains similar information for the case of four fair coins.

TABLE 1. The number of cases and the probability for each outcome *r heads showing*, when two, three, or four fair coins are flipped.

Number of Coins Flipped	Number r of Heads Showing	Number of Cases Having r Heads Showing	Probability of Having r Heads Showing
Two	0	1	1/4
	1	2	2/4
	2	1	1/4
Three	0	1	1/8
	1	3	3/8
	2	3	3/8
	3	1	1/8
Four	0	1	1/16
	1	4	4/16
	2	6	6/16
	3	4	4/16
	4	1	1/16

The numbers which appear in the third column of Table 1 are called *binomial coefficients*. Binomial coefficients are among the most widely occurring numbers in all of mathematics. We will now develop a method for computing binomial coefficients in general.

DEFINITION. Suppose n coins are to be flipped. By the symbol

$$\binom{n}{r}$$

(read "n choose r"), we mean the number of different ways in which the outcome *r heads showing* can arise.

For example, we saw that if three coins are to be flipped, there are three different ways in which we can obtain a single head showing. Therefore,

$$\binom{3}{1} = 3.$$

THEOREM 1. For each value of r from 1 to $n-1$, we have

$$\binom{n}{r} = \frac{n(n-1)(n-2)\cdots(n-r+1)}{r!} = \frac{n!}{r!(n-r)!}.$$

Proof. We will use a method of counting the number of different cases of r heads showing which contains some "overkill," and then at the end we will compensate for the overestimate. Imagine the n coins numbered 1 through n. Corresponding to each way in which the n coins can show r heads, there is a different way of choosing r numbers from the numbers 1 through n—namely, choose those numbers whose coins lie heads up. Therefore, counting the number of ways in which r heads can turn up in flipping n coins is the same as counting the number of ways in which we can pick r different numbers from the collection $1, 2, 3, \cdots, n$.

To do the latter, we might imagine that we have a long, thin container with r spaces (like an egg carton with only one row), which we proceed to fill with numbers chosen from the collection $1, 2, 3, \cdots, n$. Starting with the container empty, we have n choices of what to put in the first space. For the second space, having used up one number already, we have $n-1$ choices of what to put in it. At the third space we have $n-2$ choices left, and so on. At the last space, the rth, $r-1$ of the numbers $1, 2, 3, \cdots, n$ have already been used, so the number of choices for that space is

$$n - (r-1) = n - r + 1.$$

This filling-in process is illustrated in Figure 4 for the special case $n = 6$ and $r = 4$.

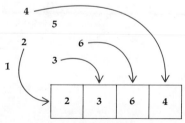

FIGURE 4. Illustration of a choice of four numbers from the collection 1, 2, 3, 4, 5, 6, to be placed in order in an "egg-carton." The total number of ways of accomplishing this is $6 \cdot 5 \cdot 4 \cdot 3 = 360$.

To find the total number of ways of filling in the r spaces, we take the product of the number of choices at each step:

$$n(n-1)(n-2) \cdots (n-r+1).$$

Now this number overestimates the number in which we are interested, because it counts the number of ways in which r of the numbers $1, 2, 3, \cdots, n$ can be put into the r positions *in order*. Given any r of the numbers $1, 2, 3, \cdots, n$, there are $r!$ ways of ordering them. (We see this by the argument we just gave: take $n = r$.) Thus the number

$$n(n-1)(n-2) \cdots (n-r+1)$$

was obtained by counting each collection of r numbers $r!$ distinct times; it is $r!$ times larger than the number in which we are interested.

Therefore, to obtain the number we want, all we need to do is divide. The number of ways in which one can choose r different numbers from the collection $1, 2, 3, \cdots, n$ is equal to

$$\frac{n(n-1)(n-2) \cdots (n-r+1)}{r!}. \tag{1}$$

This proves the first equality of the theorem. Since the numerator of (1) can be written as $n!/(n-r)!$, the second equality in the theorem is also true. This completes the proof.

Notice that Theorem 1 holds for all possible values of r except the lowest ($r = 0$) and the highest ($r = n$). The only reason the theorem does not hold in these two cases is that in these cases $0!$ would appear in the formula for $\binom{n}{r}$, and so far we have defined *factorial* only for natural numbers. If we set $0! = 1$, then Theorem 1 holds even for $r = 0$ and $r = n$. Accordingly, most mathematicians have agreed to define $0!$ to be equal to the number 1. Let us also agree to this definition of $0!$, so we may regard Theorem 1 as holding for all values of r from 0 to n.

CONSEQUENCE TO THEOREM 1. $\binom{n}{n-r} = \binom{n}{r}.$

Proof.

$$\binom{n}{n-r} = \frac{n!}{(n-r)!(n-(n-r))!} = \frac{n!}{(n-r)!r!} = \binom{n}{r}.$$

Now that we have such a nice pair of formulas for computing $\binom{n}{r}$, you may find yourself forgetting the original definition in favor of one or the other of the formulas. Whatever you do, though, you should remember the following result.

THEOREM 2. In flipping n fair coins, the probability of obtaining exactly r heads is

$$\binom{n}{r}\frac{1}{2^n}.$$

Proof. In flipping one coin, there are two possible ways in which the coin can land; in flipping two coins, there are four possible ways in which the coins can land; in flipping three coins, there are eight possible ways in which the coins can land. In general, for each additional coin flipped, the number of ways in which the coins can land doubles, so that if n coins are flipped, there are 2^n possible ways.

If all n coins are fair, each of these 2^n possibilities is equally likely, so the probability of any one of them is $1/2^n$. Exactly $\binom{n}{r}$ of these ways will have r heads showing. Therefore, the probability of obtaining exactly r heads by flipping n fair coins is

$$\binom{n}{r}\frac{1}{2^n}.$$

This completes the proof.

EXERCISES

1. Draw a tree diagram similar to Figures 2 and 3 for the act of flipping four coins. Count the number of cases showing 0 heads, 1 head, 2 heads, 3 heads, and 4 heads.

2. List the 4 ways of picking one number from the collection 1, 2, 3, 4. List the 6 ways of picking two numbers from the collection 1, 2, 3, 4. List the 4 ways of picking three numbers from the collection 1, 2, 3, 4. List the 1 way of picking four numbers from the collection 1, 2, 3, 4. [*Hint:* You can make use of the tree diagram in Exercise 1. To complete the picture, notice that there is 1 way of "picking" zero numbers from the collection 1, 2, 3, 4. That is, do not pick any.]

3. Use either formula in Theorem 1 to compute $\binom{n}{r}$ for $n = 1, 2, 3, 4, 5$, and 6 and all values of r from 0 to n.

4. Use Theorem 2 to calculate the probability of obtaining exactly k heads when $n = 2k$ fair coins are flipped, for $k = 1, 2, 3, 4$, and 5. (Notice that the probability *decreases* as k increases; this behavior is surprising to many people, who think there is a "law of averages" which says this probability should increase as more coins are flipped.)

5. Repeat Exercise 4 for $k = 10$, 15, and 20.

6. Suppose you are in the business of designing 3-color flags for 100 different countries. You wish to design the flags so that no two countries will be given the same triple of colors. (For instance, only one country should be a red-white-and-blue country.) What is the smallest number of colors you will need to produce the hundred flags? (Do you know, offhand, how many countries presently have red-white-and-blue flags?)

7. On a sheet of graph paper plot the values of $\binom{10}{r}$ versus the values of r. Draw a smooth curve through the 11 points you have plotted. Repeat this procedure for $\binom{15}{r}$, drawing a smooth curve through the resulting 16 points. What features would you say the two curves have in common?

8. Show that
$$\binom{n}{0} + \binom{n}{1} + \binom{n}{2} + \cdots + \binom{n}{n-1} + \binom{n}{n} = 2^n.$$
[*Hint:* The only really easy way to do this problem is to make use of the original definition of $\binom{n}{r}$.]

9. Show that
$$\binom{n}{r} = \binom{n}{n-r}$$
directly from the definition of $\binom{n}{r}$. [*Hint:* If r heads are showing on n coins, how many tails are showing?]

10. Let n be a fixed number. For what value(s) of r is $\binom{n}{r}$ the largest? Prove your answer if you obtained it by guessing. [*Hint:* Note that $\binom{n}{r+1} = \frac{n-r}{r+1}\binom{n}{r}$.]

11. Show that
$$\binom{n-1}{r-1} + \binom{n-1}{r} = \binom{n}{r}.$$
[*Hint:* Count the number of ways of obtaining r heads if Coin 1 shows a head and then count the number of ways of obtaining r heads if Coin 1 shows a tail.]

12. Show that if $(x+y)^n$ is multiplied out, then the coefficient of $x^r y^{n-r}$ is $\binom{n}{r}$. (This is the reason the numbers $\binom{n}{r}$ are called binomial coefficients.)

3. A Central Limit Theorem

For values of n less than 50, it is a reasonable (although sometimes not pleasant) task to calculate any number $\binom{n}{r}$ and from it the probability of obtaining r heads when n fair coins are flipped. But when the values of n become larger—and in practical problems they often be-become much larger—it is necessary to use an approximation technique. This technique is furnished by what is called a *central limit theorem*. The somewhat mysterious name of the theorem refers to its great importance *(central)* and to its increasing usefulness and accuracy *(limit)* the larger n becomes.

The particular central limit theorem we state here was first proved by the English mathematician Abraham de Moivre (1667–1754) in the year 1733. It is the simplest of all such theorems, but even so its proof requires ideas from calculus. We state it in very pragmatic terms.

THEOREM 3. For n at least as large as 20 and for r less than $n/2$, the probability of obtaining r or fewer heads when n fair coins are flipped can be determined to a good approximation simply by knowledge of the number

$$x = \frac{(n-1-2r)^2}{n}.$$

This probability can be found by taking one-half the number opposite the value x in Table A of the Appendix.

Let us concentrate our attention on the "magic number" x, so as to acquire some idea of what Theorem 3 means. Obtaining r or fewer heads means obtaining 0 or 1 or 2 or $\cdots$ or r heads. All $r+1$ of these outcomes are *at least as extreme* as the outcome r heads, in the sense that they all are at least as far away from the "even split" outcome or outcomes ($n/2$ heads if n is even and $(n-1)/2$ and $(n+1)/2$ heads if n is odd) as the outcome r heads is. To the other side of the "even split" outcomes are $r+1$ more cases that are *at least as extreme* as the outcome r heads: 0 or 1 or 2 or $\cdots$ or r tails.

3. A CENTRAL LIMIT THEOREM

Thus, there are altogether $2r + 2$ outcomes at least as extreme as r heads. Since the total number of outcomes, 0 heads through n heads, is $n + 1$, the number of outcomes *less extreme* than r heads is

$$n + 1 - 2r - 2 = n - 1 - 2r.$$

(See Figure 5 for an illustration of the less-extreme and the at-least-as-extreme outcomes for 2 heads out of 10 coins.)

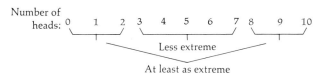

FIGURE 5. The cases at least as extreme as 2 heads out of 10 coins and the cases less extreme than 2 heads out of 10 coins.

Therefore, we have found an interpretation for the numerator of our number x; it is the *square* of the number of outcomes less extreme than the outcome r heads. It is very reasonable that the number of outcomes less extreme than r heads should be linked to the probability of obtaining r or fewer heads, because that number is in some sense a measure of just how extreme the outcome r heads is.

But all by itself, the number $n - 1 - 2r$ is not a sufficiently good measure of extremeness. For example, if 10 coins were tossed and no heads obtained, we would be rather surprised. But if 100 coins were tossed and 45 heads obtained, we would not be so surprised, because 45 is proportionally much closer to .50 than 0 is to 5. Yet in each case $n - 1 - 2r$ has the same value, 9

It turns out that dividing the *square of* $n - 1 - 2r$ by the number n produces just the right measure of extremeness. As long as n is 20 or larger, the probability of obtaining r or fewer heads is determined to a good approximation by $(n - 1 - 2r)^2/n$ alone. Once we know this number, we can find out the probability by consulting Table A of the Appendix. No knowledge of n or r individually is required.

We should give some explanation of why Table A lists double the probabilities in which Theorem 3 is interested. The reason is that the probability of obtaining an outcome at least as extreme as r heads is exactly twice the probability of obtaining r or fewer heads. (The probability of obtaining r or fewer tails is equal to the probability of obtaining r or fewer heads. These two groups of outcomes together comprise all the outcomes which are at least as extreme as the outcome r heads.) Now, as we have seen from the discussion following Theorem 3, it is natural to consider outcomes that are at least as extreme as the outcome r heads. As we continue our study of probability

204 CHAP. 10 INTRODUCTION TO PROBABILITY THEORY

and statistics, it almost always will be most natural to find the probability of obtaining an at-least-as-extreme-as outcome, and so Table A has been constructed to fit in with what most frequently will be sought from it.

After this long-winded discussion, let us do a sample problem of the sort which led us to consider Theorem 3 in the first place. Once we get our feet back on the ground by successfully completing the problem, it might be a good idea for you to reread the earlier discussion to see how it fits into a practical problem.

For 75 fair coins, we will calculate the probabilities of obtaining 30 or fewer heads and 29 or fewer heads. From these two probabilities, we will calculate the probability of obtaining exactly 30 heads.

Letting $n = 75$ and $r = 30$, we find that our measure of extremeness is

$$\frac{(75 - 1 - 60)^2}{75} = 2.613.$$

From Table A, then, we find that the probability of obtaining an outcome at least as extreme as 30 heads is .1060. One-half this number, .0530, is the probability of obtaining 30 or fewer heads.

Letting $n = 75$ and $r = 29$, we find that our measure of extremeness is

$$\frac{(75 - 1 - 58)^2}{75} = 3.413.$$

From Table A, then, we find that the probability of obtaining an outcome at least as extreme as 29 heads is .0647. One-half this number, .0324, is the probability of obtaining 29 or fewer heads.

Subtracting, we find that the probability of obtaining exactly 30 heads is

$$.0530 - .0324 = .0206$$

This completes the example.

To summarize this section, we have developed a practical method, somewhat involved, which enables us to calculate probabilities associated with flipping a large number of fair coins. It may look as though the theory of coin flipping is still the goal, and that our method of measuring extremeness is merely a means to that goal. Not so. What we really wanted was the method and Table A. The study of coin flipping was the quickest, simplest way to develop the method. Now, just as it happened in the history of probability theory, the gambling aspect will begin to recede as we turn our thoughts to ideas that are much more important.

EXERCISES

1. Using Theorem 3, find the probability of obtaining at most 22 heads when 60 fair coins are flipped. Find the probability of obtaining at most 21 heads when 60 fair coins are flipped.

2. Using Exercise 1, find the probability of obtaining exactly 22 heads when 60 fair coins are flipped.

3. Find the probability of obtaining at least 38 heads when 60 fair coins are flipped. Find the probability of obtaining at least 39 heads when 60 fair coins are flipped. Find the probability of obtaining exactly 38 heads when 60 fair coins are flipped. [*Hint:* Before you sit down and start calculating, compare this exercise with Exercises 1 and 2.]

4. Find the probabilities of obtaining exactly 19, 20, 21, 23, 24, and 25 heads when 60 fair coins are flipped.

5. Suppose 10 fair coins are flipped. Using Theorem 2, calculate for $r = 0, 1, 2, \cdots, 9$, and 10 the probabilities of obtaining exactly r heads.

6. Use Theorem 3 to calculate approximations to the answers of Exercise 5. Compare the two sets of answers. [*Note:* Although the approximations are fairly good even for such a small value of n, they generally are not considered good enough until n is as large as 20.]

7. If you know how to use logarithms, use them and Theorem 2 to calculate the probability of obtaining exactly 22 heads when 60 fair coins are flipped. Compare your answer with the answer to Exercise 2.

8. For $k = 5, 10, 15, 20$, and 30, use Theorem 3 to calculate the probability of obtaining exactly k heads when $2k$ fair coins are flipped. Compare your first four answers with the answers to Exercises 4 and 5 of the preceding section. (Note once again that the probability decreases the larger k becomes.)

4. Introduction to Hypothesis Testing

Suppose you are walking along the streets of a city, and you happen to see a car bearing the license number 3. You very likely would remark to yourself what an unusual occurrence that is. However, seeing the license number 3 is no more unusual than seeing the license number 845,229 — assuming all cars get the same exposure to passers-by — because in a given state there is only one car bearing a given license number.

In what sense, then, is the sight of the license number 3 unusual? First, it belongs to a certain small collection of low numbers (and the number 845,229 does not). Second, you have, long before seeing the license number 3, somehow decided to be amazed whenever you see a number in that small collection, which for example might consist of the numbers 1 to 100. You are, of course, justified in being amazed, for it is an unlikely event to see a number from that small group out of perhaps 1,000,000 license plates that could come by.

This simple idea of agreeing when to be amazed at an event is the key to one of the most important uses of mathematical statistics today, the testing of hypotheses.

We will illustrate hypothesis testing with two examples. The test used in these two illustrations is too simple to be very useful in practical applications. However, it is still a good idea to pay close attention, because the very simplicity will enable you to grasp the underlying ideas common to this test and the more complicated, more useful tests which appear in the next chapter.

The first example concerns U.S. pennies. In an experiment to detect possible imbalance in pennies, 25 pennies were balanced on edge on a smooth, level-as-possible table top. They were balanced facing in random directions to counter the effects of any minute slope of the table top. The table was then pounded with gradually increasing force until the vibrations had toppled all the pennies. It was found that 19 of the 25 pennies were lying heads up. The question we ask is, "How remarkable is this result if pennies are just as likely to topple one way as the other?"

Here, just as in the case of the license numbers, we must have an understanding of what we mean by *remarkable*. Guided by our thoughts on the license numbers, we see that we will have to set up a certain class of numbers such that any outcome of r heads, r belonging to that class, will be considered unusual. Our discussion of extreme outcomes in Section 3 furnishes a natural way to do this. In the case of 25 perfectly balanced coins, the two most extreme outcomes are 0 heads and 25 heads. The two next most extreme outcomes are 1 head and 24 heads. And so forth. All the outcomes at least as extreme as the outcome 19 heads are shown in Figure 6.

We pick a small number, .05 being one highly honored by tradition, and we decide to include the outcomes r heads and r tails in our class of unusual outcomes if the probability of obtaining an outcome

4. INTRODUCTION TO HYPOTHESIS TESTING

at least as extreme as these, under the assumption that pennies are all fair, is at most .05. This produces a certain class of numbers r, coming in from the two extremes of 0 and 25, and we agree to be amazed if the number 19 falls into this class.

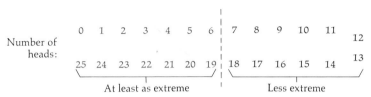

FIGURE 6. The outcomes at least as extreme as 19 heads out of 25 pennies and the outcomes less extreme than 19 heads out of 25 pennies.

Amazed at what? First, we are amazed that we should have obtained such an extreme result if pennies were really fair. But, second, we turn this amazement into amazement—even disbelief—that pennies are fair, since we *did* observe the result of 19 heads. That is, if the number 19 falls into the class of unusual numbers, we *reject the hypothesis* of fairness, because under that hypothesis the outcome 19 heads is unusual.

How do we find out if 19 belongs to the class of unusual numbers? We do *not* need to find out precisely what this class consists of. All we need to do is see if 19 meets the membership requirement. Is the probability of obtaining an outcome at least as extreme as 19 heads, assuming pennies to be fair, at most equal to .05? To find out, we view 19 heads as 6 tails and then apply Theorem 3 with $n = 25$ and $r = 6$. We find

$$x = \frac{(25 - 1 - 12)^2}{25}$$
$$= 5.76.$$

Therefore, from Table A we find that the probability of obtaining a result at least as extreme as 19 heads under the assumption of fairness is .0164. Therefore the outcome 19 heads does belong to the class of unusual cases, and so we *reject* the hypothesis of fairness.

Some comments are now in order. In performing any hypothesis test, one must choose some low number, like our .05, to be used as an upper limit on a probability. This number is used to set off a certain collection of outcomes as ones that are unusual if some hypothesis, like our assumption that pennies are balanced, is true. This number is known as a *significance level*.

Any hypothesis test involves a computation made under a certain assumption, such as the one that pennies are balanced. That assumption is known as the *null hypothesis* for the test. If a person finds that:

(1) the probability of obtaining an outcome at least as extreme as the actual experimental result
(2) is less than the chosen significance level
(3) under the assumption that the null hypothesis is true,

then he is "amazed" at the null hypothesis. That is, he *rejects* the null hypothesis.

As we will see, hypothesis testing fits in beautifully with many experimental techniques. Often a person is interested in concluding that a certain relationship is true, and he is able to set up an experiment and statistical test with the opposite of the suspected relationship as the null hypothesis. If he can reject the null hypothesis, he will then be concluding that the suspected relationship is true.

Notice that two things contribute to the decision to accept or reject the null hypothesis. The one is the actual experimental result, and the other is the choice of significance level. The lower the significance level, the fewer are the "unusual" outcomes that lead to the decision to reject the null hypothesis. The higher the significance level, the greater is the number of cases in which the null hypothesis is rejected. If the scientist is at all prejudiced in favor of rejecting the null hypothesis (and usually that is his position), once he has seen his experimental results he will be tempted to choose a significance level leading to rejection of the null hypothesis. Such a choice, however, *defeats the purpose* of hypothesis testing. Therefore, it is *very important* that one choose the significance level *before* he becomes acquainted with the experimental results.

Choice of significance level is a matter of taste, the two most popular ones today being .05 and .01. The level .05 is the less conservative of the two. It will lead to rejection of the null hypothesis more often than will the level .01, but at the same time it gives less protection against rejecting a null hypothesis that actually is true.

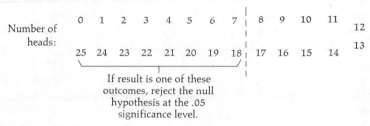

FIGURE 7. The collection of outcomes unusual at the .05 significance level, for 25 fair coins.

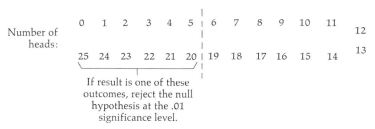

FIGURE 8. The collection of outcomes unusual at the .01 significance level, for 25 fair coins.

The differences between the two significance levels are illustrated for our 25 pennies in Figures 7 and 8. Since a significance level must be chosen before experimental results are seen, let us make our choice early in the game. In *all* tests performed in this book, let us choose .05 as our significance level.

We will now proceed to our second example of a hypothesis test. This test is occasionally used in practical work. It is often called the *sign test*, due to the use of plus and minus signs in setting it up. Suppose you are interested in seeing whether a certain drug (for example, an antihistamine) has any effect on a person's reaction time. One way of testing this suspected relationship is to perform an experiment in which each one of a number of subjects is tested twice. You would measure each subject's reaction time once with the drug and once without the drug. Suppose, for example, you obtain for 20 subjects the reaction times (in seconds) given in Table 2. You decide that if an individual's time is lower with the drug than without, he will be awarded a plus sign, while if his time is higher with the drug, he will be awarded a minus sign. (The use of the two signs + and − is from tradition; they do not represent value judgments. The letters H and T or any pair of different symbols would serve as well.) We see that you have obtained 16 minuses and 4 plusses.

The assumption that the drug has no effect on reaction time is the null hypothesis. Under this assumption, each person should have been just as likely to receive a + as to receive a −. That is, each person should have behaved exactly like a fair coin with sides + and −. Therefore, we can test the null hypothesis in exactly the same way as we did in the penny example. Here we have $n = 20$ and $r = 4$. The value of x is

$$\frac{(20 - 1 - 8)^2}{20} = 6.05.$$

Therefore, from Table A we find that under the null hypothesis the probability of obtaining a result at least as extreme as 4 plusses out of

20 trials is .0139. Since this probability is less than the agreed-upon significance level, the experimental result belongs to the collection of unusual outcomes, so we reject the null hypothesis.

TABLE 2. Reaction times of 20 subjects with and without antihistamine dosage (data fictitious).

Subject	Reaction Time without Drug	Reaction Time with Drug	Sign
A	1.05	1.10	−
B	.95	.93	+
C	.45	.65	−
D	.75	.80	−
E	.61	.70	−
F	.83	1.05	−
G	.50	.63	−
H	.47	.50	−
I	.87	.85	+
J	.98	1.31	−
K	.95	.97	−
L	1.00	1.13	−
M	.91	.89	+
N	1.20	1.10	+
O	.77	.81	−
P	.73	.74	−
Q	.81	1.10	−
R	.60	.65	−
S	.47	.93	−
T	.72	.85	−

This simple test can be useful in experimental work where one is able to test each subject more than once and under different conditions. However, when the ability to retest subjects does not exist, it is difficult to get those all-important plusses and minuses. Therefore, the test can only be used on data taken under controlled conditions that allow for retesting each subject. You can regard it as our first example of a statistical test that can (occasionally) be used in practice. However, it is serving us primarily as a precursor of the three much more useful tests to be discussed in the next chapter.

Let us conclude this section with a few more comments about penny flipping. In 1967, as a byproduct of a research project I assigned in an elementary probability course, I acquired a rather large sample of the behavior of pennies. The pennies were "flipped" by shaking them ten at a time in a drinking glass and then spilling them out into a box lid. The result was 160,136 heads out of 319,020 penny flips.

That is, heads turned up 50.196 percent of the time. At the .05 significance level, with the null hypothesis that the pennies used were fair, this result belongs to the collection of unusual cases.

Having that much information, I began to wonder if unfairness of pennies would show up more dramatically when given the chance. So I devised the experiment of balancing pennies on edge. Depending to some extent on the roughness of the surface on which they are balanced, pennies topple heads up approximately $\frac{3}{4}$ of the time.

Once you look for it in a penny, you will readily find the cause of the imbalance. Almost all coins today are stamped with the engraved features lying below the rim, so that the engraving is protected from wear. The heads side of the Lincoln penny has relatively strong relief, which is accomplished by the background's being stamped deep into the coin, considerably below the rim. The tails sides on both old and new pennies have shallower relief, and so the backgrounds on them are not stamped so far below the rim. The result is that the mass of the penny is shifted toward the tails side, and so (presumably) a penny balanced on edge or falling on edge will have a tendency to topple with the tails side down.

EXERCISES

1. Suppose in toppling 25 pennies you obtain 18 heads and 7 tails. Under the assumption that pennies are balanced, what is the probability of obtaining an outcome at least as extreme as the one actually obtained? Do you accept or reject the null hypothesis at the .05 significance level?

2. Suppose that in the example of the sign test in this section we had obtained 5 plusses and 15 minuses for the 20 subjects. Under the null hypothesis (that the drug has no effect), what is the probability of obtaining a result at least as extreme as this one? Do we accept or reject the null hypothesis at the .05 significance level?

3. Fourteen U.S. pennies are set spinning rapidly. When they cease to spin, 2 are found lying heads up, and 12 are found lying tails up. State the appropriate null hypothesis and then perform a hypothesis test using the experimental evidence. (Remember that the use of Table A for n less than 20 is risky. Use Theorem 2 instead.)

4. Under the assumption that the pennies used are fair, calculate the probability of obtaining a result at least as extreme as 160,136 heads out of 319,020 flips.

212   CHAP. 10   INTRODUCTION TO PROBABILITY THEORY

5. You have flipped 17 coins, obtaining 4 heads. If you test the null hypothesis of fairness using Theorem 2, do you accept or reject the null hypothesis? If you test by use of Table A, do you accept or reject the null hypothesis?

6. Suppose you are performing the sign test on n subjects, all of whom have received a plus sign. What values of n are too small to enable you to reject the null hypothesis at the .05 significance level?

7. Verify that the collections of unusual cases in Figures 7 and 8 are correct.

8. Perform your own experiment toppling approximately 30 pennies which you have balanced on edge. Carry out the hypothesis test to see if you should accept or reject the null hypothesis of fairness.

9. Flip a coin (any coin) 100 times, keeping track of how many times it turns up heads. Test the null hypothesis that the coin is fair.

10. If 40 students do Exercise 9, the probability is high (.8715 approximately) that at least 1 student will reject the null hypothesis even if all 40 coins used are fair. Explain why this is so. (You need not attempt to compute the number .8715; a qualitative explanation is sufficient.) Of all the students in your class who did Exercise 9, how many did reject the null hypothesis?

11. Suppose we flip n coins. For $n = 10^2, 10^3, 10^4, 10^5, 10^6$, calculate the smallest number r of heads greater than $n/2$ which is needed in order to reject the null hypothesis of fairness at the .05 significance level. In each of the five cases, calculate the number

$$p_n = \frac{r}{n} - \frac{1}{2}.$$

From your five values of p_n, can you guess the (approximate) behavior of p_n as n increases?

5. The Notion of a Test Statistic

In the two examples of the previous section, you have seen the basic procedure for hypothesis testing. It involves:

5. THE NOTION OF A TEST STATISTIC

(1) setting up a null hypothesis,
(2) choosing a significance level,
(3) seeing if the experimental result belongs to a collection of results that are unusual if the null hypothesis is true, and
(4) deciding to reject or accept the null hypothesis.

In order to perform Step (3), it is often convenient to compute from the experimental data a number, such as

$$x = \frac{(n-1-2r)^2}{n}.$$

Tabled values of this number then enable us to determine what the probability is of obtaining a result at least as extreme as the one actually obtained, under the assumption that the null hypothesis is true. If this probability is no more than the significance level (always .05 for us), then the result belongs to the collection of unusual results under the null hypothesis, and so we reject the null hypothesis. Otherwise, we accept the null hypothesis.

A number such as x, which plays such an important role in the process of deciding to reject or accept the null hypothesis, deserves a special name. It is called a *test statistic*.

If one always uses the same significance level, he will need to consult a given table only once, even though he makes many decisions of acceptance or rejection. The reason is that he can tell from the value of the test statistic alone whether or not the probability of an at-least-as-extreme result will fall below the agreed-upon significance level. For example, in the two problems we worked in the last section, we could have rejected the null hypothesis merely by comparing the test statistic to *one number*.

What number? Let us consult Table A to find out. We see that for values of x less than 3.84, the corresponding probabilities are all greater than .05, while for values of x greater than or equal to 3.84, the corresponding probabilities are all less than or equal to .05. It is in the second case that we reject the null hypothesis at the .05 significance level. That is, we can reject the null hypothesis if and only if we have a value of x at least as large as 3.84. The number 3.84 is called the *critical value* for the test statistic x.

To repeat, as long as we are performing hypothesis tests at the .05 significance level, the only piece of information we need from Table A in order to accept or reject the null hypothesis is the critical value 3.84. However, the exact value of the probability associated with a given value of the test statistic is worth knowing in itself, so we will continue to use the full Table A.

EXERCISES

1. Using Table A, find the critical values of the test statistic x for significance levels .10, .02, .01, and .005.

2. Since the use of Table A is not justified for fewer than 20 trials, we cannot use the fraction

$$x = \frac{(n - 1 - 2r)^2}{n}$$

as a test statistic when n is less than 20. Instead, we can use the number r or $n - r$, whichever is smaller. For all values of n less than 20, use Theorem 2 to compute the critical value of this test statistic for the .05 significance level.

CHAPTER ELEVEN

Three Statistical Tests

1. Introduction

In this chapter you will be outfitted with a very useful kit of three statistical tests. They can be used to solve a large fraction of the commonly occurring problems in hypothesis testing. Each of the three is probably (in the author's opinion) the best of its kind available. Far from being cute examples or toys, they are what the experts in the field use.

The subject of statistics has a reputation for the variety of ways in which people have misused its techniques. It is possible to misuse any statistical test ever invented. And it is a fact that most misuses arise from someone's having learned a test without having learned as well the common sense limitations that must be observed in applying it. Bearing this in mind, we will devote the next section to studying the role of statistical tests in research. Also, in the sections that follow it, we will frequently comment on possible pitfalls and even point out weaknesses in our own examples. You, the reader, are encouraged to raise further questions and objections, above and beyond those raised in the text.

Nearly every application of a statistical test will suffer from one or more minor weaknesses. An expert avoids making statistical arguments having serious weaknesses, but as a matter of practicality he is willing to live with some minor weaknesses. If he were not, he would almost never perform a statistical test.

To sharpen your eye for pitfalls and have fun at the same time, you may be interested in reading *How to Lie with Statistics* by Darrell Huff. Some of Huff's examples apply strictly to presentation of data in charts and so on, while others demonstrate pitfalls that must be avoided in applying hypothesis tests. From a very practical point of view, it is better for a person to be weak on statistical formulas (which he can always look up in a book) and strong on knowing when it is dangerous to use the formulas than it is for him to be strong on formulas but oblivious to pitfalls.

2. The Role of Statistical Tests in Research

Hypothesis testing in research has a very definite place, and that place is near the end of the research process. You may have already acquired that idea from the nature of the hypothesis test. It does exactly what its name suggests; it tests hypotheses. And before a person can test hypotheses, he must already have put in enough thought and investigation to have formulated the hypotheses. Also, before he can test hypotheses, he will need to have gathered data with which to test them. A typical outline of research aided by statistical tests is given below.

(1) Thought and investigation lead to formulation of a theory.

(2) From the theory, one or more relations are predicted to hold among the members of certain groups. As much as possible, the relations are chosen to be ones that will not be predicted by other theories. The assumption that a predicted relation does *not* hold is called a *null hypothesis*.

(3) Random samples are taken from the groups in which the relations have been predicted to hold.

(4) Statistical tests are performed to see if the sample results are *unusual* under the assumption that the null hypothesis is true.

(5) If the sample results are unusual (according to a previously agreed-upon significance level), the null hypothesis is rejected. The theory is then considered to be true (in a pragmatic sense), because it has led to one or more correct predictions.

With regard to Step (5), if the null hypothesis is not rejected, one does not automatically discard the theory or the predicted relation. It may well have happened that the relation is present but weak, so that a larger sample is needed to detect it. At present, we cannot be precise about the role that the size of the sample plays.

Once we have developed the three tests, we will then be in a position to study their behavior with different sample sizes. However, it seems reasonable that additional information in the form of more samples should enable one to detect more of the weaker relations that might hold.

You now have a fairly good idea of the indirectness of hypothesis testing. Hypothesis testing is definitely not a proof or disproof in the sense of a mathematical proof. It does, however, fit in very well with the modern view or philosophy of science. So many theories in the sciences have held sway for a time, only to be replaced by more advanced ones, that everyone has given up hope of ever knowing the full "truth" about nature or reality.

Rather, successive theories are regarded now as being better and better descriptions of reality, where the goodness (or truth) of a description is measured by: (1) its agreement with observation and (2) its ability to make predictions that can be verified by experiment or further observations. There are examples of this theory succession in practically all sciences, but the classic example is the replacement of Newton's theory of mechanics, which dates from the seventeenth century, by the two theories of quantum mechanics and relativity theory, both of our own century. And no one doubts that these theories will one day be replaced by better ones yet.

3. The Pearson Test: Testing for a Relation between Two Conditions

Frequently it is conjectured that within a group there is a relation between two different conditions. Perhaps the most widely publicized conjecture of this sort today is that there is a relation (say, in the population of the United States) between the condition of being a tobacco smoker and the condition of having lung cancer.

Notice that the relation proposed is statistical in nature. Not every tobacco smoker has lung cancer, nor is every lung cancer patient a tobacco smoker. Yet, from the mass of sample data available, it has been possible to reject the null hypothesis (that tobacco smoking and having lung cancer are unrelated), and thus conclude that there is a relation between the two conditions. What, precisely, is that relation? The relation is that in the collection of people from which the sample was drawn (say, the population of the United States), the group of smokers contains a *higher proportion* of lung cancer victims than does the group of nonsmokers.

Notice that nothing has been said about cause and effect. A theory which predicts that smoking and lung cancer are related conditions may well have involved the reasoning that tar and nicotine stimulate cancerous growths, and the decision that the relation is present will thus be evidence in favor of the theory. However, a simple decision that a relation exists between two conditions does *not* necessarily mean that one condition is the cause of the other.

First of all, one condition may be a cause and the other an effect, but the statistical test alone will never tell a person which is which. Second, it may happen that both conditions are effects of a single cause and thus turn out to be related, even though neither is a cause of the other. In the chapter "Post Hoc Rides Again" of *How to Lie with Statistics*, you will find several examples of people noticing a relation between two effects of the same cause and then falsely concluding that one effect is the cause of the other. The most spectacular of these is the conclusion that drinking milk causes cancer. (!)

To illustrate the test for a relation between two conditions, we will examine a relation tested by a student of the author in a statistics course term paper. The student's theory was that young people who felt they had a close family life would tend to be more enthusiastic about raising families of their own. To test his theory, he asked the following three questions of his sample group:

(1) Do you consider your family closer than most?
(2) How many brothers and sisters do you have?
(3) How many children would you like to have?

The relation he predicted was that in a population of roughly the same age (say students at a certain kind of university), those answering yes to Question 1 would have a higher proportion wanting at least the same size of family than would those answering no to Question 1. His sample group consisted of the students within his statistics course at Northwestern University. Lacking a more truly random sample of students, he gave his reasons for believing his sample was fairly representative of students at schools like Northwestern. The results obtained from his sample are shown in Table 1.

Table 1 is known as a *contingency table*. From it we can see that the predicted relation is present among the members of the sample. That is, those answering yes to Question 1 have a higher proportion wanting at least the same size family than do those answering no to Question 1. The question we now ask is, could this difference in proportions have occurred by chance if the predicted relation does *not* hold in the population from which the sample was drawn? The assumption that the relation does not hold in the population is the

3. THE PEARSON TEST

null hypothesis for the test we are about to perform. To answer the question, we compute a test statistic which we call p:

$$p = \frac{N(|BC - AD| - N/2)^2}{(A + B)(C + D)(A + C)(B + D)}$$

$$= \frac{120(|1334 - 360| - 60)^2}{73 \cdot 47 \cdot 38 \cdot 82}$$

$$= 9.38.$$

TABLE 1. Contingency table showing relation of closeness of family life to desire for same or larger sized family. The sample consists of 120 Northwestern University students.

Feel that Immediate Family Is Closer than Most

	NO	YES	
YES	15 = A	58 = B	$N = A + B + C + D$ = 120
NO	23 = C	24 = D	

Wish to Have at Least as Many Children as Are in Present Family

The statistic p is one of a very general class of test statistics discovered by Karl Pearson (1857–1936). The most important part of the formula for p is the portion $|BC - AD|$. (The vertical lines mean that the sign in front of the number $BC - AD$ is to be dropped.) This number is a measure of the contingency table's departure from proportionality, because the closer $|BC - AD|$ is to 0, the closer the equation $BC = AD$ is to holding, and this latter equation can be rewritten as

$$\frac{B}{A} = \frac{D}{C}$$

provided A and C are larger than 0. The additional term $N/2$ is a correction term; usually it is small compared with $|BC - AD|$, as was true above.

Thus, the statistic p is a measure of how close the contingency table is to being proportional. If p is small, the table is close to proportional, while if p is large, the table is not close to proportional. That is, under the null hypothesis, the nonextreme values of p are those close to 0, while the extreme values are those farther from 0.

It turns out that if the sample size N is at least 40, and if the null hypothesis is true, the probability that p will be at least as large as the number x is determined (to a good approximation) by the value of x alone. It also turns out that this probability is given by Table A of the Appendix once again.

Therefore, at the .05 significance level, the critical value for the statistic p is 3.84. Since 9.38 is larger than the critical value 3.84, we reject the null hypothesis at the .05 significance level. (Consulting Table A, we find that the probability of obtaining a value of p at least as extreme as 9.38 if the null hypothesis is true is approximately .0022.) The student's theory is thus supported by the existence of the relation he predicted.

Recall that we observed from the contingency table that the relation in the sample was in the direction predicted by the student's theory. That is, the ratio $B/A = 58/15$ was larger than the ratio $D/C = 24/23$. It is possible to tell the direction of the relation without computing these ratios. Simply look at the sign of $BC - AD$. Positive sign means that B/A is larger than D/C; in this case, we say that the two conditions are *positively correlated* in the sample. (For example, we would say that in the sample of 120 Northwestern students, the two conditions of feeling one's immediate family is closer than most and wishing to have at least as many children as are in one's immediate family are positively correlated.) Negative sign means that B/A is less than D/C; in this case we say that the two conditions are *negatively correlated* in the sample. Should A or C be equal to 0, one or the other of the ratios B/A and D/C will not exist, but we still speak of positive or negative correlation within the sample according to whether the sign of $BC - AD$ is positive or negative. Very rarely it will happen that $BC - AD = 0$, in which case we say that there is *no correlation* within the sample.

Notice that positive or negative correlation in a sample means nothing for the population from which the sample is drawn *unless* the correlation is strong enough (as measured by the statistic p) that it enables us to reject the null hypothesis.

Having finished our discussion of the Pearson test as applied to the student's example, we should now spend some time discussing possible noncomputational shortcomings of the study. There do not seem to be any serious flaws, but in the business of statistics a person learns to be cautious, lest someday a flaw overlooked turn out to invalidate an entire study.

A major consideration always is, "How well does the sample represent the population from which it is drawn?" The assumption which underlies almost every statistical test, including this one, is that the sample has been drawn from some population in a totally random fashion. The word *random* means that every member of the population stood exactly the same chance of being picked to be in the sample—like lottery tickets drawn from a thoroughly mixed barrel or cards dealt from a thoroughly shuffled deck. From the point of view of statistics, a random sample is the most representative kind of sample there is, because each member of the population stood equal chance of being chosen to be in the sample.

Now, it is quite possible that family sizes, present and desired, may be different for students at Northwestern and similar universities than for students at, say, state universities. It is also possible that the proportion of students believing their families to be closer than most will vary with the kind of university. Therefore, it seems necessary to consider the sample to be representative only of institutions similar to Northwestern.

However, detection of a predicted relation in any population can be taken as evidence in favor of the theory, *provided* there is no reason why the relation should hold in that particular population if it did not hold elsewhere. If there were a reason, that reason would constitute another theory predicting the same relation for that population. In that case, we would have to admit to having chosen a sample that would not allow us to distinguish between two possible explanations. However, after a bit of soul-searching we do not see any reason why the relation should hold among students at Northwestern and similar schools if it does not hold elsewhere. Therefore we can regard the demonstration of the relation in the one population as good support of the original theory.

Now let us turn our attention to another worry. Suppose closeness of family life tended to be negatively related to the size of the subject's present family. Suppose also that, regardless of closeness of family life, the subjects all tended to want the same number of children. Then the relation predicted by the student would hold, but *not* because students from close families were especially enthusiastic about wanting families of their own. Rather, it would hold because the nonclose families tended to be the larger families, and students from these, in wishing for the same number of children as the other students wished for, were thus wishing for smaller families.

222 CHAP. 11 THREE STATISTICAL TESTS

Thus, there is a simple competing theory which predicts the same relation as did our original theory. However, it is easy to knock this second theory down, just by looking for the additional relation which it presumes—that the students from larger families tend to feel their family life is not close. Table 2 shows the distribution of the 120 Northwestern students' families with respect to closeness and number of children. Although we presently do not have the technique to perform a statistical test with these data, it is apparent from sight alone that there is little difference in the sample between close and nonclose families.

TABLE 2. Distribution of the 120 students' families with respect to closeness and number of children.

	Number of Children in Present Family							
	1	2	3	4	5	6	7	8 or more
Close Families	IIII	JHT JHT JHT JHT III	JHT JHT JHT JHT JHT	JHT IIII	JHT I		II	IIII
Nonclose Families		JHT JHT II	JHT JHT JHT III	JHT JHT II	I		II	II

We will close this section with another example—and a puzzle. Consider Table 3, which shows the distribution of white U.S. citizens born in 1966 with respect to two conditions, sex and time of year child was born (middle six months versus the three months on each end).

The first observation we make is that for U.S. children born in 1966, being male is positively correlated with being born in the middle six months of the year. The second observation we make is that the contingency table is very near to being proportional, much more so than Table 1. But our third observation is that p for Table 3 is large—over 7. (The exact value is left to you to compute as an exercise.) We remark, then, that if the sample size is large, a slight disproportionality in the contingency table can lead to a large value of p and thus to rejection of the null hypothesis. Loosely speaking, the larger the sample size the weaker is the weakest relation that can be detected.

Very large sample sizes are regarded by statisticians as mixed blessings. They make it possible to detect minute differences in proportions. However, if a difference is so very small, it may be of academic interest only and therefore not worth detecting. Also, the test statistic p for large samples is so sensitive to slight proportional changes in the data that the tiniest departure from randomness in the sample can send it skyrocketing.

Regarding the positive correlation in Table 3, then, one would be interested in seeing it in other years and in seeing it for nonwhites as well as whites. I have checked, and it does seem to hold in general.

There is no theory that I know of to explain the correlation. I just happened to notice it while looking for something else in the vital statistics.[1] The reason for dividing up the months as was done was simply because the mid-year months were the months in which the male-to-female birth ratio was higher. The computation of p was done mainly as a matter of curiosity, to see if the difference in proportions could be due to chance alone. The answer seems to be a resounding no. Although the difference is far too small to be of practical use, I confess I am still curious why there should be any difference at all.

TABLE 3. Contingency table showing relation of child's sex to time of birth. The sample consists of all white children born in the United States in 1966.

		NO	YES
Child Is Male	YES	756,248 = A	779,238 = B
	NO	720,274 = C	737,470 = D

Child Born in Middle Six Months: April through September

[1] Others have noticed this correlation also. For a number of comments and observations on human sex ratios, see I. M. Lerner. *Heredity, Evolution, and Society*. San Francisco: W. H. Freeman and Company, 1968, pp. 120–123.

EXERCISES

1. Compute p for Table 3.

2. Compute p for the contingency table each of whose entries is twice that of Table 3. Compare your answer with the answer to Exercise 1.

3. Compute p for the contingency table each of whose entries is twice that of Table 1. Compare your answer with the value of p, 9.38, computed from Table 1 in the text.

4. The sample giving rise to Table 3 consists of all white children born in the United States in 1966. Of what population might this sample be a representative sample? Give a reason for your answer.

In Exercises 5 to 14, you will be given theories and data taken to test those theories. In each exercise you should (a) state an appropriate null hypothesis, (b) perform the Pearson test, (c) tell whether the correlation between the two conditions is positive or negative in the sample, and (d) list any flaws that might cause the exercise to be a misuse (rather than a use) of statistics. The symbol $YY:8$ means that 8 people answered yes to both questions. The symbol $YN:16$ means that 16 people answered yes to the first question and no to the second question, and so on.

5. Theory: Beer drinking and swinging go together. Questions: Are you a steady consumer of beer? Are you a virgin? Answers: $YY:8$, $YN:16$, $NY:60$, $NN:35$. Sample: 119 students in statistics course at Northwestern University.

6. Theory: Women are more looks-conscious than men. Questions: Are you a male? Do you wear frame-type glasses? Answers: $YY:36$, $YN:27$, $NY:21$, $NN:36$. Sample: 120 students in statistics course at Northwestern University.

7. Theory: Fraternity-sorority membership and clothes-consciousness are related. Questions: Do you belong to a Greek organization? Do you think professors who wear the same sweater to class all the time look all right? Answers: $YY:20$, $YN:35$, $NY:23$, $NN:45$. Sample: 123 students in statistics course at Northwestern University.

8. Theory: Those feeling less secure about their good looks tend to be more concerned about dress. Questions: Are you above-

average in looks? Do you think professors who wear the same sweater to class all the time look all right? Answers: *YY*:32, *YN*:52, *NY*:11, *NN*:26. Sample: 121 students in statistics course at Northwestern University.

9. Theory: At certain universities, first impressions can be deceiving. Questions: Are you a freshman? Do you think the students here are very friendly? Answers: *YY*:68, *YN*:40, *NY*:19, *NN*:58. Sample: 185 students in topics-in-mathematics course, first day of class.

10. Theory: Supply one yourself. Questions: Do you like to dance? Do you have brown eyes? Answers: *YY*:61, *YN*:48, *NY*:29, *NN*:47. Sample: Same as in Exercise 9.

11. Theory: Women are more oriented to marriage than are men. Questions: Are you a male? Do you want to get married within the next five years? Answers: *YY*:22, *YN*:41, *NY*:37, *NN*:18. Sample: 118 students in statistics course at Northwestern University.

12. Theory: Men are the ones who are really in favor of women's liberation. Questions: Are you a male? Assuming that women are becoming more aggressive sexually, do you disapprove? Answers: *YY*:19, *YN*:46, *NY*:31, *NN*:27. Sample: 123 students in statistics course at Northwestern University.

13. Theory: In the United States, emotional forms of self-expression are considered nonmasculine. Questions: Are you a male? Do you like to sing? Answers: *YY*:41, *YN*:56, *NY*:57, *NN*:31. Sample: Same as in Exercise 9.

14. Theory: Students who are the oldest or only children in the family would rather be with other people than by themselves. Questions: Are you the oldest or only child in your family? Do you prefer to live in a room with roommates rather than in a single? Answers: *YY*:28, *YN*:13, *NY*:42, *NN*:35. Sample: 118 students in statistics course at Northwestern University.

4. The Wilcoxon Test: Testing for a Relation between a Condition and a Rating

Often a question of interest in research will have not simply a yes-or-no answer but a numerical answer that can vary over a considerable range. Examples are: What is your height? How many years have you lived at your present address? How many days has it been since your last haircut? A theory may predict that the answer to such a question will be related to the answer to some yes-no question.

There are two principal kinds of relation possible. In the first kind, those answering yes to the yes-no question will tend as a group to give either larger answers or smaller answers to the numerical answer question than will those answering no. In the second kind, those answering yes to the yes-no question will tend as a group to give numerical answers spread either over a larger range or over a smaller range than the answers given by those answering no.

In this section we will study a very useful test which is designed to detect relations of the first kind. It is sensitive to both kinds of relation, but it is far more sensitive to the first than to the second. The test was invented by Frank Wilcoxon in 1945 and two years later was studied and developed further by Mann and Whitney. (Accordingly, it is sometimes called the Mann-Whitney-Wilcoxon test or the Mann-Whitney test.) The approach we will use here was developed by Mann and Whitney.

Our first example is a fictitious one involving small samples. Suppose you are a teacher interested in comparing two methods of teaching a foreign language. In the same semester, you had taught a class of seven students by Method A and a class of nine students by Method B. At the end of the semester you measured their proficiency in the language by means of a common final examination. Suppose their scores on the examination are those given in Table 4.

TABLE 4. Final examination scores of seven students taught a foreign language by Method A and nine students taught the same foreign language by Method B (data fictitious).

Students Taught by Method A	Students Taught by Method B
100	98
93	95
89	88
87	84
86	82
70	71
55	65
	63
	40

The theory, of course, is that there is some sort of difference in the effects of the two methods, and the relation predicted is that there

4. THE WILCOXON TEST   227

will be some sort of difference in the final examination scores of students taught by each method. The null hypothesis is that there is no difference. The sample consists of the seven students taught by the one method and the nine students taught by the other method.

The first step in performing the Wilcoxon test is to rank all the individuals together, in the order of their scores. We let the letter A represent anyone taught by Method A and the letter B represent anyone taught by Method B.

lowest   B A B B A B B B A A B A A B B A   highest

It is not necessary to write down the scores themselves; one of the remarkable features of the Wilcoxon test is that it uses only the *ranking* which the scores produce.

Having lined up all the individuals, we now count the total number of times an individual from the second group (B) is found to the right of an individual from the first group (A). This number will always be designated by the letter U. The easiest way to calculate U is to count for each A the number of B's to the right of it and then sum all of these numbers. In our example, we have

$$U = 8 + 6 + 3 + 3 + 2 + 2 + 0$$
$$= 24$$

found as a sum of seven numbers, one for each A.

In the Wilcoxon test, the very smallest value that U can take is 0, which arises when all individuals in the second group (the group B in our example) are to the left of all the individuals in the first group (the group A in our example). The very largest value U can take is the size m of the first group (7 in our example) times the size n of the second group (9 in our example). This value arises when all individuals in the second group are to the right of all the individuals in the first group. Thus, in our example, the value $U = 24$ is one of the lower values which could have occurred, since the possible values ranged from 0 to 63.

Under the null hypothesis, the extreme values of U are those near either end of the range of possible values of U—out toward 0 or mn. The nonextreme values are the ones around the midway point between 0 and mn, those near $mn/2$ ($= 31.5$ in our example).

In order to obtain a test statistic whose nonextreme values are close to 0, we will define the number V to be the difference between U and $mn/2$:

$$V = U - \frac{mn}{2}.$$

In our example, then,
$$V = 24 - 31.5$$
$$= -7.5.$$

Under the null hypothesis, the extreme values of V are located near $mn/2$ and $-mn/2$. For our test statistic in this first example, we will use the size $|V|$ of V; that is, our test statistic will be V without the sign preceding it:

$$|V| = 7.5.$$

For small values of m and n, it has been possible to compute for various values of x the probability under the null hypothesis that $|V|$ will be at least as large as x. This is done by counting how many different orderings of the two samples give rise to each value of $|V|$. Since this probability depends not only on x but also on m and n, the resulting table of probabilities is huge. However, for any given significance level (like our chosen .05 level), all that is necessary for deciding to reject or accept the null hypothesis is knowledge of the critical value of the test statistic.

Critical values of the test statistic $|V|$ are given in Table B of the Appendix for all values of m and n less than or equal to 20. To use Table B, we locate the number opposite the appropriate values of m and n and then check to see if our value of $|V|$ is at least as large as the tabled number. If it is, then our value $|V|$ belongs to the class of values that are unusual under the null hypothesis, and so we reject the null hypothesis.

In our example of the language students, we find from Table B that the critical value for $m = 7$ and $n = 9$ is 19.5. Since our value $|V| = 7.5$ is less than this, it is not an unusual value under the null hypothesis. Therefore we accept the null hypothesis.

In deciding whether to reject the null hypothesis, we made use only of the size of V. Now let us ask what use might be made of the sign of V. We recall that the possible values of U range from 0 to mn, and the corresponding values of V range from $-mn/2$ to $mn/2$. The values of V with negative sign correspond to the smaller values of U and thus to orderings in which individuals of the second sample tend to occur to the left of individuals of the first sample.

If the individuals are ordered with numerical ratings increasing to the right (as we ordered them in the example), a negative value of V thus corresponds to the individuals of the second sample tending to have lower ratings than individuals of the first sample. And a positive value of V corresponds to the individuals of the second sample tending to have higher ratings than individuals of the first sample.

4. THE WILCOXON TEST

In the case the null hypothesis is rejected, one often would like to say that not only do the two populations from which the samples are drawn differ, but also that they differ by having the individuals from one population tending to rate higher than individuals from the other. Such a conclusion is not strictly correct, because the Wilcoxon test is sensitive to other differences between populations as well.

However, let us think about the problem from the point of view of vindicating a theory. If the null hypothesis is rejected, there is *some* sort of difference between the two populations, as was predicted by the theory. Thus, the theory is to some extent justified by rejection of the null hypothesis alone. But the Wilcoxon test is known to be sensitive primarily to one kind of difference: members of one population tending to rank higher than members of the other population. Therefore, if the sign of V is such that the two samples are ranked in the order predicted by your theory (assuming your theory predicted an order), you can derive a stronger justification from that fact. If, however, you have rejected the null hypothesis and the sign of V indicates the samples are ranked opposite to the way your theory said they should be ranked, you might do well to think about "revising" your theory.

Let us now consider a larger example, this time with real data. The original study was performed by a student of the author. This student had heard that traditionally there has been a remarkably low rate of alcoholism among Jews. He was interested to know if that was still the case among younger-generation Americans. The guess, or prediction, was that among U.S. college students there is a relation between the condition of being Jewish and the rating consisting of the number of times a person drinks per month. His sample consisted of the students within his statistics course at Northwestern University.

The results of his sampling are given in Table 5. The number of strokes in each compartment of the table is the number of students falling into that category. The stroke method of counting, by the way, not only is an easy way to tabulate experimental or sampled data but also yields a vivid picture of the data. We notice right away from Table 5 that there is a striking difference between the two groups of students in our sample.

We notice also that there are two differences between the data here and the data in our foreign language example. The one difference is that we have much more data in this example. That is, the sample size is too large to enable us to find the critical value of V from Table B. The other difference is that there are now quite a number of *ties*, students who have the same drinking frequencies.

TABLE 5. Distribution of 123 Northwestern University students with respect to religion and number of times each student drinks per month.

	Number of Times Student Drinks per Month										
	0	1	2	3	4	5	6	7	8	9	10 or more
Jewish $m = 31$	12	4	5	4	2	3					1
Non-Jewish $n = 92$	22	14	8	15	4	5	4	3	9		8

This second difference creates no difficulty. What we will do is generalize the definition of U very slightly so that it can count the ties properly. We generalize it as follows. In case a member of the second sample (non-Jewish students in our example) is tied with a member of the first sample, we count him, her, or it as being "one-half to the right" of the member of the first sample. We then define U just as before: U equals the total number of times a member of the second sample appears to the right of a member of the first sample.

The ties make computation of U quite easy, because much of the counting can be accomplished by a few multiplications. For example, the total number of times a non-Jewish student appears to the right of a Jewish student who drinks 4 times per month is the product

$$2 \times (8 + 9 + 3 + 4 + 5 + \tfrac{1}{2} \times 4) = 2 \times 31$$
$$= 62.$$

To find U, we sum these products over those rating categories that contain at least one member of the first sample:

$$\begin{aligned}
U =\ & 12 \times (8 + 9 + 3 + 4 + 5 + 4 + 15 + 8 + 14 + \tfrac{1}{2} \times 22) \\
& + 4 \times (8 + 9 + 3 + 4 + 5 + 4 + 15 + 8 + \tfrac{1}{2} \times 14) \\
& + 5 \times (8 + 9 + 3 + 4 + 5 + 4 + 15 + \tfrac{1}{2} \times 8) \\
& + 4 \times (8 + 9 + 3 + 4 + 5 + 4 + \tfrac{1}{2} \times 15) \\
& + 2 \times (8 + 9 + 3 + 4 + 5 + \tfrac{1}{2} \times 4) \\
& + 3 \times (8 + 9 + 3 + 4 + \tfrac{1}{2} \times 5) \\
& + 1 \times \tfrac{1}{2} \times 8 \\
=\ & 972 + 252 + 260 + 162 + 62 + 79.5 + 4 \\
=\ & 1791.5.
\end{aligned}$$

4. THE WILCOXON TEST

We then compute V just as before:

$$V = U - \frac{mn}{2}$$
$$= 1791.5 - \frac{31 \times 92}{2}$$
$$= 365.5.$$

For sample sizes to which Table B of the Appendix applies, we would of course use $|V|$ as the test statistic. For sample sizes larger than those covered by Table B, it turns out that the number

$$w = \frac{12V^2}{mn(m+n+1)}$$

is a test statistic such that under the null hypothesis the probability of obtaining a value of w at least as large as x is given (to a good approximation) by Table A. (By this time you may be wondering what black magic makes Table A appear so frequently. Some comment, but not a full answer, will be given at the end of the chapter.)

In our example, we have

$$w = \frac{12 \times 365.5^2}{31 \times 92 \times 124}$$
$$= 4.53.$$

Comparing this value with the critical value 3.84, we conclude that we should reject the null hypothesis at the .05 significance level. From Table A, we see that under the null hypothesis the probability of obtaining a value of w at least as great as 4.53 is .0333.

Now that we have rejected the null hypothesis, let us examine the sign of V. You will recall that it was positive, indicating that the second sample (non-Jewish students) tended to have higher drinking frequencies than did the first sample (Jewish students). This difference is in the direction predicted by the theory that the alcoholism rate among the younger generation of Jews in this country is still low. Therefore, we have every reason to consider our rejection of the null hypothesis a confirmation of the theory.

Let us end this section with a discussion of possible flaws in the student's research project. The first question is, "How well do the two samples represent the populations from which they are drawn?" Just as in the Pearson test, completely random samples are the ideal. One might wonder if the drawing from a particular kind of school could introduce some biases. For example, it could be that Jewish students tend to choose Northwestern for its image of academic excellence, while at least some non-Jewish students tend to choose Northwestern for its image as a socializing school. Should this be the case, then the Jewish sample could be biased in the non-drinking, serious student direction, while the non-Jewish sample could be biased in the drinking student direction.

It is my own guess that these biases, if they exist at all, are not very strong. However, the sure way to settle the difficulty would be to sample some university that is not simultaneously famous for parties and academics. In general, if you are able to obtain a sample under different circumstances, you should make use of the opportunity. Biases can creep into a sample in so many subtle ways that a second sample is always a worthwhile defensive measure.

In looking around for weaknesses, we might run across the following theory which predicts the same relation: Members of fraternities and sororities tend to lead an active party life and thus might tend to drink more often than other students do. Also, members of fraternities and sororities at most schools tend to be non-Jewish. (How much this is due to fraternities and sororities discriminating against Jews, and how much it is due to Jews not caring for the fraternity-sorority way of life will not be discussed here.) Thus, the difference between Jewish and non-Jewish students' drinking habits could be due to the fraternity-sorority set within the group of non-Jewish students.

Looked at carefully, however, this theory is not a rival to the original theory; rather, it is an amplification of the original theory. Previously it was suggested simply that a religious difference might be connected with a difference in drinking, while now it is suggested that this connection could in part be due to a tendency of Jews not to belong to certain organizations that foster drinking. This amplification is well worth exploring, but we have no space for it in the text. In the exercises, you will be asked to examine several relations having to do with the new, more detailed theory.

EXERCISES

1. Perform the Wilcoxon test on the data given in Table 2 of Section 3.

2. The format of this exercise is the same as that of Exercises 5 to 14 of the preceding section. Answer the same questions you were required to answer for those exercises. Theory: Jewish students tend to join fraternities and sororities less frequently than do non-Jewish students. Questions: Are you Jewish? Do you belong to a Greek organization? Answers: $YY:11$, $YN:23$, $NY:44$, $NN:45$. Sample: 123 students in statistics course at Northwestern University.

4. THE WILCOXON TEST

3. Perform the Wilcoxon test using the data in the table below. (Each number is the difference in inches between the waist measurement and the chest measurement of a Northwestern student.)

Male	Female
7	8
7	10.5
7.5	10.5
7.5	10.5
8	11
9	12
9	12
10	12.5
16	14

In Exercises 4 to 11, you will be given theories and data taken to test those theories. In each exercise you should (a) state an appropriate null hypothesis, (b) perform the Wilcoxon test, (c) tell which sample tended to have the higher ratings, and (d) list any flaws that might cause the exercise to be a misuse (rather than a use) of statistics. In parentheses following the rating question, you will find the possible answers to the question. The symbol Yes: 3, 0, 6, · · · means that of those who answered yes to the condition question, 3 fell into the lowest rating category, 0 into the next lowest category, 6 into the next, and so on.

4. Theory: Belonging to a fraternity or sorority is related to frequency of drinking. Questions: Do you belong to a Greek organization? How many times per month do you drink (0, 1, 2, 3, 4, 5, 6, 7, 8, 9, 10-or-more)? Answers: Yes: 11, 7, 4, 7, 3, 6, 2, 2, 8, 0, 5. No: 23, 11, 9, 12, 3, 2, 2, 1, 1, 0, 4. Sample: 123 students in statistics course at Northwestern University.

5. Theory: We are interested in seeing whether the relation between religion and drinking is still present in the part of our sample that is not "Greek." Questions: Are you Jewish? How many times per month do you drink (0, 1, 2, 3, 4, 5, 6, 7, 8, 9, 10-or-more)? Answers: Yes: 10, 3, 4, 2, 2, 1, 0, 0, 0, 0, 1. No: 13, 8, 5, 10, 1, 1, 2, 1, 1, 0, 3. Sample: 68 non-Greek students in statistics course at Northwestern University.

6. Theory: In college, women are more socially oriented than men. Questions: Are you a male? How many dates have you had in

the last seven weeks (0, 1, 2, 3, 4, 5, 6, 7, 8, 9, 10, 11, 12, 13, 14, 15, 16, 17, 18, 19, 20, 21-or-more)? Answers: Yes: 7, 3, 6, 10, 6, 9, 4, 1, 1, 0, 4, 0, 2, 1, 4, 1, 0, 0, 0, 0, 2, 2. No: 3, 1, 0, 7, 2, 2, 5, 1, 3, 2, 4, 4, 4, 1, 3, 2, 1, 0, 0, 0, 2, 12. Sample: 122 students in statistics course at Northwestern University.

7. Theory: Strength of religious convictions is related to wealth. Questions: Do you think praying has any effect? What, to the nearest thousand dollars, was the income of your parents last year (0–9, 10–14, 15–19, 20–24, 25–29, 30–49, 50-or-more)? Answers: Yes: 12, 21, 12, 8, 4, 8, 5. No: 4, 9, 2, 4, 5, 7, 4. Sample: 104 students in statistics course at Northwestern University.

8. Theory: Those who would demand a rigid obedience of children will also tend to collect a large number of acquaintances for use as "admirers." Questions: Do you agree, obedience and respect for authority are the most important virtues children should learn? At Northwestern, how many good close friends would you say that you have, that you are with often or talk to often (0, 1, 2, 3, 4, 5, 6, 7, 8, 9, 10, 11–15, 16–20, 21-or-more)? Answers: Yes: 0, 0, 2, 1, 4, 4, 3, 2, 2, 1, 4, 1, 3, 3. No: 3, 4, 5, 13, 17, 12, 8, 3, 5, 0, 10, 6, 5, 1. Sample: 122 students in statistics course at Northwestern University.

9. Theory: A dormitory is not a good place to study. Questions: Do you usually study in the dormitory? What was your grade point average for the fall quarter (0, .75, 1.5, 1.75, 2, 2.2, 2.25, 2.3, 2.5, 2.7, 2.75, 2.8, 3, 3.25, 3.3, 3.5, 3.7, 3.75, 4)? Answers: Yes: 0, 1, 0, 2, 5, 0, 6, 2, 7, 3, 8, 0, 10, 5, 1, 2, 0, 5, 2. No: 1, 0, 1, 1, 4, 1, 1, 1, 7, 3, 7, 2, 8, 9, 1, 10, 2, 3, 1. Sample: 122 students in statistics course at Northwestern University.

10. Theory: Men are more politically aware than are women. Question: Are you a male? Rating: Score on test concerning current political figures (0, 1, 2, 3, 4, 5, 6, 7, 8, 9, 10, 11, 12, 13, 14, 15, 16). Answers and scores: Yes: 1, 0, 3, 5, 4, 7, 6, 4, 5, 6, 13, 1, 1, 1, 1, 1, 1. No: 1, 4, 6, 5, 10, 6, 6, 6, 1, 9, 3, 0, 0, 0, 0, 0, 0. Sample: 117 students in statistics course at Northwestern University. Sample question from test: Which of the following states did Richard Nixon win last November [1968]: New York, Pennsylvania, Ohio, Illinois, Texas, California? The test was given in February 1969.

11. Theory: Students with close family ties will tend to have a higher regard for traditional institutions, values, and beliefs. Question: Do you consider your family closer than most? Rating: Score on

tradition respect test (0, 1, 2, 3, 4, 5, 6, 7, 8, 9, 10, 11, 12, 13, 14, 15, 16, 17, 18, 19, 20, 21, 22, 23, 24, 25, 26). Answers and scores: Yes: 0, 0, 0, 0, 0, 1, 1, 1, 0, 7, 0, 3, 1, 6, 3, 3, 9, 2, 4, 7, 6, 3, 3, 2, 2, 0, 0. No: 0, 0, 0, 1, 0, 0, 3, 3, 2, 2, 2, 3, 3, 5, 1, 1, 1, 5, 1, 0, 4, 3, 2, 0, 0, 0, 0. Sample: 106 students in statistics course at Northwestern University. Sample questions from test: Have you been in a demonstration? Do you agree, what this country needs now, more than laws, are leaders in whom the people can place their trust? If you agreed with the goals of a student demonstration, would you take an active part in it?

5. The Kendall Test: Testing for a Relation between Two Ratings

In the previous section we developed a test which is used to study a *single* rating in relation to a condition. However, it often happens in research that two rating scales are either developed or present themselves for attention, and it is suspected that they may be related. Perhaps it is suspected that a person tending to score high on one rating scale will tend to score high on the other. For example, students scoring high on CEEB tests usually go on to achieve high grade point averages in college. (Were this not the case, the CEEB presumably would go out of business.)

Or a high score on one scale might tend to go with a low score on another. For example, members of the United States Senate who score high on the ADA (Americans for Democratic Action) scale usually score low on the ACA (Americans for Constitutional Action) scale.

These two kinds of relations are the ones to which the Kendall test is primarily sensitive, but to a lesser extent it is sensitive to other relations which might hold between two ratings.

Our first example of the use of Kendall's test is a fictitious one involving small samples. Suppose I have devised a test to predict a person's success as an automobile salesman, and I want to know if it really works. I administer the test to eight prospective salesmen A, B, C, D, E, F, G, and H. Suppose they score in the order:

lowest $C \ E \ A \ F \ H \ D \ B \ G$ highest

Then I turn them loose in comparable automobile dealerships, and one year later I find that they have made the profits given (in hundreds of dollars) in Table 6.

TABLE 6. Profits in hundreds of dollars for eight apprentice automobile salesmen (data fictitious).

C	E	A	F	H	D	B	G
153	136	145	160	197	185	300	217

Notice that the individuals are placed in the table in order of their scores on the first rating, the test. For each individual i, I perform the following subtraction:

$M_i =$ the number of profits higher than i's profit and to the right of i *minus* the number of profits lower than i's profit and to the right of i.

The results of these subtractions are:

$$M_C = 5 - 2 = 3 \qquad M_H = 2 - 1 = 1$$
$$M_E = 6 - 0 = 6 \qquad M_D = 2 - 0 = 2$$
$$M_A = 5 - 0 = 5 \qquad M_B = 0 - 1 = -1$$
$$M_F = 4 - 0 = 4 \qquad M_G = 0 - 0 = 0.$$

Then I calculate the sum S of all these M_i:

$$S = 3 + 6 + 5 + 4 + 1 + 2 - 1 + 0$$
$$= 20.$$

The number S is a measure of how strongly the second rating (profit in our example) is related to the first rating (test score in our example). Had the eight salesmen's profits increased continuously from left to right, we would have had for our value of S:

$$S = 7 + 6 + 5 + 4 + 3 + 2 + 1 + 0$$
$$= 28$$

and this is the largest value of S possible with a sample size of 8, since each M_i is as large as possible.

Thus, the maximum value of S corresponds to perfect agreement between the two ratings. The minimum possible value of S is -28, the negative of the maximum. This value would have been obtained had the profits decreased continuously from left to right. Under the null hypothesis (no relation between the two ratings) the values 28 and -28 and values close to them are the extreme values. The values midway between them, close to 0, are the nonextreme values.

That is, S is playing a role here exactly like that of the number V in the Wilcoxon test. We can therefore make a pretty safe guess as to what will be a useful test statistic: $|S|$. For samples of size less than or equal to 40, the critical values of this statistic for the .05 significance level are given in Table C of the Appendix.

Looking up sample size 8 in Table C, we find the critical value to be 18. Since our value of $|S|$ is 20, we see that it belongs to the class of results which are unusual if the null hypothesis is true. Therefore we reject the null hypothesis at the .05 significance level.

Now that we have concluded there is some sort of relation between the two rating scales, we would like to say that they are related in such a way that the higher a person ranks on the first scale (the test), the higher his rank will tend to be on the second (profit) scale. Strictly speaking, such a conclusion is not correct, but since the Kendall test is primarily sensitive to this relation and to its opposite, once we reject the null hypothesis, we are fairly safe in concluding that one or the other relation holds in the population from which the sample is drawn. Since our extreme result $S = 20$ is positive, the ratings of the eight individuals in the sample tended to agree with each other. Thus, we are fairly safe in saying that the two ratings will tend to agree with each other in the population from which the sample was drawn.

In general, if we obtain a positive value of S, we will say that the two ratings are *positively correlated* within our sample, and if we obtain a negative value of S, we will say that the two ratings are *negatively correlated* within our sample. If the size of S is large enough that we reject the null hypothesis, then we are fairly safe also in presuming that the two ratings are similarly correlated in the population from which the sample was drawn.

Let us now consider a larger sample, this time with real data. It was conjectured that the income of a male student's parents would be related to the amount of social life he has in college. The reasoning behind this guess is not profound; students from less well-to-do families often work part-time and so have less time as well as less money to spend on pleasurable activities. It was predicted that the student's rating on a parental income scale would be related to his rating on a second scale—the number of dates he had had in the previous seven weeks.

The 55 male students in a statistics course at Northwestern University were asked to estimate the number of dates they had had in the previous seven weeks and to estimate into which of seven categories the income of their parents fell the previous year. Their answers are given in Table 7, where the seven columns represent the different income categories. Within each column is the number of dates reported by each student in that parental income category.

TABLE 7. Numbers of dates in a seven-week period and income category of parents for 55 Northwestern University male students.

Parental Income in Thousands of Dollars						
0 to 9	10 to 14	15 to 19	20 to 24	25 to 29	30 to 49	50 and up
1	0	0	0	2	0	6
3	0	4	1	5	2	10
3	0	5	3	14	2	12
4	0	5	5	21	3	14
4	1	5	10		4	30
5	2	7	10		6	
6	2		12		20	
20	2		14			
	3					
	3					
	3					
	3					
	4					
	4					
	5					
	8					
	15					

As we did in the small sample, for each individual i we compute

$M_i = $ the number of students in columns to the right of i who had more dates than i *minus*
the number of students in columns to the right of i who had fewer dates than i.

For example, consider any one of the three entries "5" in the 15,000 to 19,000 column. The numbers in the columns to the right which are greater than it are: 10, 10, 12, 14, 14, 21, 6, 20, 6, 10, 12, 14, and 30, thirteen in all. The numbers in the columns to the right which are less than it are: 0, 1, 3, 2, 0, 2, 2, 3, and 4, nine in all. Thus, for each of these entries "5," we have the value of M_i:

$$M_i = 13 - 9$$
$$= 4.$$

The entries in individual i's column are not counted, nor are the *ties* in other columns counted. Thus ties in the data turn out to be no problem at all. We ignore them completely.

Once all the M_i have been calculated, we compute their sum S, finding (in our example)

$$S = 300.$$

5. THE KENDALL TEST

Since S is positive, we know that in our sample the number of dates is positively correlated with parental income. Now we must answer the question, "Is a value of $S = 300$ unusual if the null hypothesis holds for the population from which our sample was drawn?"

For a sample size of 40 or under, we would of course use the critical values of S in Table C to answer this question. For sample sizes larger than 40, we compute the number

$$k = \frac{18S^2}{n(n-1)(2n+5)}.$$

Under the null hypothesis, the probability of obtaining a value of k at least as large as the number x is given (to a good approximation) by Table A of the Appendix.

For our values $S = 300$ and $n = 55$, we find that

$$k = \frac{18 \cdot 300^2}{55 \cdot 54 \cdot 115}$$
$$= 4.74.$$

Comparing this value with the critical value of 3.84, we reject the null hypothesis at the .05 significance level. From Table A, we find that under the null hypothesis the probability of obtaining a value of k at least as large as 4.74 is .0295.

We have concluded that, in the population from which our sample was drawn (say, U.S. male college students), there is a relation between parental income and the number of dates in a given period. We have one more thing to do. Our theory suggests that this relation ought to consist of a positive correlation, so we should look to see which kind of correlation we had in our sample. The value of S, 300, was positive, so the correlation in the sample was positive, in agreement with what was expected.

Let us end this section by considering a single, rather serious difficulty with the study we have just completed. The answers to both questions, parental income and number of dates, were obtained from the students themselves. The answer to the parental income question undoubtedly involved some guesswork on the part of the student, and the answer to the dating question involved the student's own definition of what constituted a date. It is possible that a tendency to overestimate the one answer could be coupled to a tendency to overestimate the other answer. Were this the case, it could account for a part or all of the positive correlation in the sample.

The problem of how truthfully people answer questions, even when they are assured of anonymity, is a very serious one for a statistician. Often it is difficult or even impossible to obtain information without asking questions.

You might be interested in returning to the alcoholism study of Section 4, to see if lack of truthfulness could have been a worry there. Also, if you ever embark upon a research project of your own, you should weigh the advantages and disadvantages of gathering data by questions as opposed to other means. In case there is no way other than asking questions, it is often possible to phrase the questions in ways that will yield answers which are as truthful as possible.

EXERCISES

1. In the following table, the individuals A through J have been ordered according to their scores on a vision test, the lowest on the left and the highest on the right. Below each individual in the table is the number of mistakes he failed to detect in a test drawing containing thirty errors. Perform a Kendall test to see if the vision test and the error detection test are related. Is the correlation in the sample positive or negative?

B	A	G	J	I	C	E	D	F	H
3	11	7	2	0	1	0	3	4	0

2. It was once theorized (by a relative of the author) that people of small stature tend to grow their hair long as compensation for their build. Among 113 students in a Northwestern University statistics course, it was found that there *was* a very strong negative correlation between height and length of time since last haircut. Explain why this correlation, although very strong, does not necessarily support the theory. [*Hint:* The sample contained 62 men and 51 women.]

3. The table that follows gives the numbers of dates in a seven-week period and income category of the parents for 51 female Northwestern University students. Perform a Kendall test to see if parental income and frequency of dating are related in the population of U.S. female college students. Give reasons why you might not expect such a strong relation among women as among men.

Parental Income in Thousands of Dollars						
0 to 9	10 to 14	15 to 19	20 to 24	25 to 29	30 to 49	50 and up
3	0	1	5	0	8	6
5	0	3	21	3	11	12
8	4	6	21	3	12	13
16	6	7		6	12	20
	6	8		9	14	49
	9	10			20	
	10	10			28	
	11	11			49	
	11	12				
	14	15				
	30	15				
		21				
		49				
		49				
		50				

4. The table that follows gives the number of times a person drinks per month and the person's authoritarianism score for 51 female

Authoritarianism Score						
0	1	2	3	4	5	6
0	0	0	0	5	2	3
	0	0	0	5	4	
	0	0	1	8	12	
	0	0	1			
	1	0	1			
	1	1	3			
	1	1	3			
	1	2	3			
	1	2	3			
	4	2	4			
		2	4			
		3	8			
		3	8			
		3				
		3				
		3				
		3				
		8				
		8				
		10				

Northwestern University students. Perform a Kendall test to see if the two ratings are related in the population from which the sample was drawn. Can you think of any theory that would predict such a relation? (The authoritarianism score was taken to be the number of yes answers to the following seven assertions: (1) What this country needs now, more than laws, are dedicated leaders in whom the people can place their trust. (2) What a person does is not so important as that he does it well. (3) Young people often get rebellious ideas, but as they grow up they have to get them over with and settle down. (4) Discipline and determination invariably make the difference between a successful life and a failure. (5) Most people should have a belief in a supernatural power whose decisions they obey without question. (6) If people would talk less and work more, everyone would be better off. (7) Obedience and respect for authority are the most important virtues children should learn.)

5. The following table gives the sum of a person's two CEEB scores together with the number of children in the person's family for

Number of Children in Family								
1	2	3	4	5	6	7	8	9 or more
1112	1100	1117	1100	1074	940	1120	1350	1026
1450	1100	1127	1100	1100		1151		1100
	1150	1140	1308	1143		1210		1180
	1175	1195	1309	1200		1400		1206
	1180	1263	1350	1200				1215
	1250	1271	1399					
	1281	1295	1411					
	1300	1310						
	1300	1325						
	1320	1330						
	1362	1342						
	1367	1346						
	1375	1350						
	1378	1400						
	1400	1425						
	1439	1440						
		1448						
		1472						
		1473						
		1531						

61 male Northwestern University students. Perform a Kendall test to see if the two ratings are related in the population from which the sample was drawn. Can you think of any theory which would predict such a relation?

6. (For those familiar with Chapter 7.) Explain the connection between the method used to determine whether a rearrangement is even or odd and the method used in this chapter to compute S.

6. Table A and Sample Sizes

You have now seen Table A arise in four different places, and you should be getting curious about its seeming universality. There is no one reason why Table A arises everywhere it does. Various central limit theorems have been proved showing that the Table A probabilities arise under very general circumstances, but even the most general of these theorems do not cover all the cases in which the table is used.

There is one circumstance, however, which is common to nearly all appearances of Table A. That is that the table gives *approximations* to probabilities, and that these approximations become more accurate the larger a certain number n (usually sample size) becomes.

The official name of Table A is "Table of Probabilities for a Chi-Square Random Variable with One Degree of Freedom." If you look in other books on statistics, you will not find such an extensive table of Chi-Square. Rather, you will find the equivalent information in a table entitled "Table of Probabilities for a Normal Random Variable." In the latter table, you would need to look up $\sqrt{x}$, $\sqrt{p}$, $\sqrt{w}$, and $\sqrt{k}$. Use of the Chi-Square table enables one to avoid the effort and possible errors involved in computing square roots.

Let us now ask what happens to the value of any one of the test statistics if the (total) sample size is increased from n to $2n$, but all proportions within the sample stay the same. For example, in the case of the statistic

$$x = \frac{(n - 1 - 2r)^2}{n},$$

how does x change if we flip $2n$ coins and obtain $2r$ heads (so that the proportion of heads remains the same)?

For the new value of x we would have the fraction

$$\frac{(2n - 1 - 4r)^2}{2n}.$$

Changing the number 1 to the number 2 has very little effect on the value of this fraction, but watch what the change enables us to do:

$$\frac{(2n-2-4r)^2}{2n} = \frac{4(n-1-2r)^2}{2n}$$
$$= 2\frac{(n-1-2r)^2}{n}.$$

Therefore, doubling the number of coins will approximately double the value of the test statistic x if the proportion of heads does not change.

Let us do the same with the statistic

$$p = \frac{N(|BC-AD|-N/2)^2}{(A+B)(C+D)(A+C)(B+D)}.$$

That is, let us suppose we have a sample of size $2N$, and that the entries in the contingency table are $2A$, $2B$, $2C$, and $2D$, as in Table 8.

TABLE 8. Contingency table with doubled sample size but unchanged proportions.

2A	2B
2C	2D

Since the fraction $N/2$ is only a correction term and does not have much effect for reasonably large N, let us ignore it:

$$\frac{2N(2A \cdot 2D - 2B \cdot 2C)^2}{(2A+2B)(2C+2D)(2A+2C)(2B+2D)}$$
$$= \frac{32N(AD-BC)^2}{16(A+B)(C+D)(A+C)(B+D)}$$
$$= 2\frac{N(AD-BC)^2}{(A+B)(C+D)(A+C)(B+D)}.$$

Therefore, again we find that doubling the number of samples will (approximately, since we neglected $N/2$) double the value of the statistic p, provided the proportions in the contingency table do not change.

One can show similar approximate behavior for the Wilcoxon and Kendall statistics w and k. More generally, one can show that if the sample size is increased c times, and if the proportions within the sample do not change, then the value of the statistic x, p, w, or k will be increased approximately c times.

But wait just a minute here. Does this mean that, if we obtained a value $p = 2$ (which causes us to accept the null hypothesis at the .05 level), then all we need do is double the sample size to obtain $p = 4$, whereupon we can reject the null hypothesis? Not at all. If it did, we could always arrange to reject the null hypothesis by taking a large enough sample, and hypothesis testing would be sheer nonsense.

Remember that the doubling of the statistic p with sample size took place under the assumption that the new contingency table was proportional to the old. Now any contingency table or any set of data has its proportions determined partly by relations within the population from which it is drawn and partly by chance. If you were to draw ten samples from a single population, all samples the same size, and if you were to make a contingency table from each sample, the chances are very good that you would have ten different contingency tables and ten different values of p. Since we do not stand much chance of obtaining the same value of p by taking the same size sample again, we do not stand much chance of exactly doubling p by taking a sample twice as large.

However, if you are convinced that a value $p = 2$ is in large part due to the null hypothesis not holding, and if you are willing to make the effort of taking a sample *three* to *four* times as large (to allow for chance fluctuations of p), then you will stand a good chance of obtaining a value of p greater than 4 with the new sample.

We return to a comment made in Section 2. We remarked there that the acceptance (or nonrejection) of the null hypothesis does not represent nearly such a firm commitment on the part of the researcher as does a rejection of the null hypothesis. The original sample size may have been too small to give the test statistic much of a chance to detect the amount of relationship actually present.

Choosing a reasonable sample size to allow a good chance of detection can be a difficult task, unless some small amount of data is already at hand, or unless you indicate that you simply are not interested in detecting a relation whose strength (measured by some criterion) falls below a certain level. With some data on hand, the relation of the test statistic to sample size under unchanging proportions can be used very handily to make a guess at an adequate sample size. With no data on hand, the tendency of most researchers is to underestimate the sample size.

In part, cost and time always tend to influence the choice of sample size in a downward direction. The other major downward influence is the belief that mathematical statistics is so powerful a tool that it can detect relationships invisible to the naked eye in a sample. Usually, however, the reverse is true. Many a relation which looks so strong in the sample turns out to be one of the usual chance fluctuations occurring under the null hypothesis. One of the greatest services of mathematical statistics has always been deflating false conclusions made by "eyeballing" the data.

CHAPTER TWELVE

Computer Programming

1. Introduction

The digital computer is an influence in the life of practically everyone in our country today. Amazing things are being done with computers, but few people have a clear idea of what these amazing things are or how they are done. The purpose of this chapter is to give some idea of the capabilities—and the limitations—of computers.

In keeping with the philosophy of this book, you will learn how people solve problems with computers by learning for yourself how to solve some problems on a computer. That is, you will learn how computers work by actually learning how to program a computer.

It is easy to learn to program, because computers are basically simple organisms—much simpler in their "mental" processes than you or I. The main difficulty in learning to program, in fact, is one of talking over the computer's head. To be a successful programmer, a person must be able to give directions to the computer on a level that the computer can understand. He must build up his instructions from expressions that the computer has the ability to comprehend.

Fortunately, it is now much easier to make oneself understood to computers than it was in the mid-1950s. The reason is that a number of "high-level" programming languages are now available. The first of these languages was FORTRAN, which made its appearance in 1957. These languages are called "high-level" because they are much closer to the kind of language that human beings understand than they are to the kind that computers understand naturally. When a person submits a program in FORTRAN to a computer, the computer does not make direct use of the statements he has written. Instead, it translates them (usually in two steps) to its very own "machine" language and then follows the translated instructions.

FORTRAN was developed by a team of IBM scientists headed by John Backus (1925–   ). The project was begun in 1954 and took three years to complete. The introduction of FORTRAN in 1957 produced a revolution in computing. Naturally, the language was adopted by almost all users. Since that time, two major scientific programming languages have been developed that are generally admitted to be better than FORTRAN, namely, ALGOL and PL/I, but they have yet to push FORTRAN from its top position. It seems that once people latch onto a good thing, they are not always willing to let go in order to use something better. We will follow suit; all programming discussed in this chapter will be done in FORTRAN.

2. FORTRAN Formulas

The word FORTRAN is a shortening of "formula translation." In the FORTRAN language you will be able to write algebraic formulas in much the same way you did in high school algebra and succeeding courses. The computer assumes the responsibility of translating your formulas into its own machine language.

In writing FORTRAN formulas, the main thing to remember is that statements must be given to the computer line by line, and therefore all formulas must be written out line by line. For instance, the fraction

$$\frac{X+Y}{Z}$$

written in FORTRAN will have to appear all on one line. We do have a standard method in algebra for writing that fraction on one line; namely,

$$\frac{X+Y}{Z} = (X+Y)/Z.$$

As much as possible, FORTRAN formulas follow the one-line formulas we are already used to. In fact, in this case the FORTRAN version of the above fraction is exactly the one-line form that we wrote down:

$$(X+Y)/Z$$

2. FORTRAN FORMULAS

FORTRAN allows for five arithmetic operations: addition, subtraction, multiplication, division, and raising to a power. It represents these five operations by the symbols $+$, $-$, $*$, $/$, and $**$. Table 1 shows standard algebraic expressions using each of these operations along with the FORTRAN versions.

TABLE 1. Some typical algebraic expressions together with their FORTRAN equivalents.

Algebraic	FORTRAN
$X + Y$	X + Y
$X - Y$	X − Y
XY	X * Y
$\dfrac{X}{Y}$ or X/Y	X/Y
X^Y	X ** Y
$\dfrac{(X+Y)^{XY}}{X^2 - Y}$	$((X + Y) ** (X * Y))/((X * X) - Y)$ or $((X + Y) ** (X * Y))/((X ** 2.) - Y)$
$A + B - C + D + 10.5$	A + B − C + D + 10.5

As is illustrated by the last line of Table 1, if parentheses are not needed in an algebraic formula involving addition and subtraction, then neither are they needed in the FORTRAN version of that formula. There are also some rules which enable a person to do away with some parentheses in the other operations, but in general it is safest and best to use parentheses liberally, so as to indicate clearly which operations are to be performed first. Parentheses, when used, must be balanced. That is, there must be the same number of ")" as there are of "(." Everyone agrees that this rule is important, but violations of it are common mistakes when there are many sets of parentheses.

So far we have talked about the symbols for the five operations. Now let us discuss the "things" joined by the operations. These "things" fall into two broad categories, *constants* and *variables*.

Constants are simply decimal numbers like

  123. 14147.5 −3.14159 2. 100000.

No commas are allowed in FORTRAN constants. The reason is that commas are used for another purpose and so are not free to be used within the constants. It may take us a little while to get used to 100000.

without a comma, but the computer has no such problem. *Every* FORTRAN constant must have a *decimal point*, even if the constant (such as 123.) is an integer. The comma rule and the decimal point rule are two very important things to remember.

There is a limit on the number of significant digits a FORTRAN constant can have, usually seven, eight, or nine. (A significant digit is any digit except one of the initial string of 0's.) Since this limitation is rather generous, we will never worry about it in this chapter.

A *variable* is a symbol made up (according to certain rules) of the so-called "alphameric" characters: A, B, C, · · ·, X, Y, Z (all capitals) and 0, 1, 2, 3, 4, 5, 6, 7, 8, 9. Some examples of variables are:

$$A \quad X15 \quad PAY \quad EDWARD \quad B4I4 \quad TEN$$

Just as in algebra the variable x in the formula $x^2 + x + 1$ is allowed to take on any numerical value, so too a FORTRAN variable in your program is allowed to take on any numerical value that you (via the program) tell the computer to give it. Once it has taken on a value, you can at any later stage alter that value to a new one by a suitable instruction to the computer.

Rules for writing variables are as follows:

(1) Every variable must begin with a letter, and that beginning letter must be different from I, J, K, L, M, and N.

(2) Every variable must be at most 6 characters long.

(3) Never use punctuation marks, dollar signs, and so on in variables; only letters and numerals are permitted.

(4) Do not use the following words as variables: ASSIGN, CALL, COMMON, DATA, DO, END, ENTRY, FORMAT, GO TO, PAUSE, PRINT, PUNCH, READ, REAL, RETURN, REWIND, STOP, and WRITE. These all have special meanings in FORTRAN.

Writing FORTRAN formulas with long-winded variables might seem to you to be a clumsy procedure. No one absolutely has to, because there are 20 letters of the alphabet that can be used all by themselves as variables, and usually not a great many variables are needed. However, the use of words or suggestive abbreviations as variables can be a real help to someone trying to understand a complicated program. You are free to use whatever names you wish for variables, subject only to the four rules above.

You probably have noticed that in hand printing there can be confusion between the three capital letters O, I, Z and the numbers 0, 1, 2. A computer programmer quickly picks up the habit of printing as in Figure 1 and in general prints computer-intended materials with care.

Ø   I   Z̵ (The letters)

0   1   2 (The numerals)

FIGURE 1. The usual means of distinguishing between look-alikes in FORTRAN programming.

Now that we have seen all the combinations of letters and numbers that are admitted as variables, we see the reason that the operation of multiplication cannot simply be indicated by juxtaposing two variables. The symbol AB is a single FORTRAN variable. To indicate the product of the two variables A and B, we use the asterisk as described earlier: A * B.

EXERCISES

1. Which of the following numbers are FORTRAN constants and which are not? For each one which is not, tell why it is not.

   .123 .000000000000012345 $\sqrt{2}$
   −5.6 10000 $100.
   12,526.

2. Which of the following are FORTRAN variables and which are not? For each which is not, tell why it is not.

   JACK   7TEEN   XM-148
   MARY   S   LBJ
   ALICE   GO TO   PAYROLL
   SIX12   F(10)   SYZYGY
   VOL   RAD

3. Write the following expressions as FORTRAN formulas.

   $A - (B + C)$ $(X + Y)^{X-Y}$ $P(1 + X)^Y$
   $\dfrac{A}{B + C - D}$ $\tfrac{1}{2} A \times B$ $\tfrac{1}{3} A \times B \times C$
   $4 \times 3.1416 \times R^2$

4. Write the following FORTRAN formulas in (more or less) standard algebraic notation.

   X ** (X ** X)   (X ** X) ** X
   A ** (B + (C * D))   (10. * X) * 2.
   (10. * X)/2.   (A * X)/B
   B/(A + X)   1. + X + ((X ** 2.)/2.) + ((X ** 3.)/(3. * 2.))

5. Why are not the following expressions legitimate FORTRAN formulas?

(1. − (3. + Y)/2.   A * (B + 1)
1.3 + 2.5 + X.   (A + B)(C + D)

6. Alter each of the expressions in Exercise 5 so that it becomes a FORTRAN formula.

7. Write each of the following as a FORTRAN constant.

4. ** (1./2.)   3. + 5.3 − 4.1
5. − 3.2 + .05   1./100.
.1 ** 10.   .1 * 10.

8. Write each of the FORTRAN formulas you obtained in Exercise 3 in a different form with the same meaning.

9. Write down FORTRAN formulas whose value (to the accuracy of the computer) will be

$$\sqrt{3} \quad 2^{72} \quad \sqrt[3]{4} \quad \tfrac{1}{3}.$$

3. Assignment Statements and Our First Program

In order for a computer to make use of a FORTRAN formula, it must be told to set the variables that appear in the formula equal to certain constants. Once it has been told the values of the constants, then and only then can it go ahead and combine them according to the formula. The FORTRAN statement which gives a value to a certain variable is known as an *assignment statement*. One form of this statement is

$$\text{variable} = \text{constant}.$$

For example, for a computer to evaluate the FORTRAN formula

$$(3.1416*(\text{RAD}**2.))*\text{HT} \tag{†}$$

it must be given specific values of the variables RAD and HT. We might write two assignment statements

$$\text{HT} = 10.$$
$$\text{RAD} = 3.1$$

As a result, the computer would be able to evaluate the FORTRAN formula (†) by inserting the values 10. and 3.1 for HT and RAD, respectively.

3. ASSIGNMENT STATEMENTS AND OUR FIRST PROGRAM

Suppose we *are* interested in having the computer tell us the volume of a circular cylinder whose base has radius 3.1 and whose height is 10. What we could do is feed it those two values of RAD and HT with the two assignment statements given above and then ask it to compute the volume by giving it another kind of assignment statement, namely

$$\text{VOL} = (3.1416 * (\text{RAD} ** 2.)) * \text{HT}$$

This assignment statement has the form

$$\text{variable} = \text{expression}$$

where the expression on the right-hand side is a FORTRAN formula which the computer can evaluate, since it has already been given the values of HT and RAD. Once this assignment statement has been given to the computer, it will compute VOL within a very tiny fraction of a second.

Then all we need to do is get the computer to tell us the value of VOL. That is accomplished by giving the computer an *output statement* of the form

$$\text{PRINT 999, variable}$$

In the program we are setting up, the output statement is

$$\text{PRINT 999, VOL}$$

Finally, all we need to do is finish off the program with the following statements, which we will use throughout this chapter.

$$\text{STOP}$$
$$\text{999 FORMAT (1H0, F50.20)}$$
$$\text{END}$$

Thus, the entire program for finding the volume of a circular cylinder with radius of base 3.1 and height 10. is given as follows:

$$\text{HT} = 10.$$
$$\text{RAD} = 3.1$$
$$\text{VOL} = (3.1416 * (\text{RAD} ** 2.)) * \text{HT}$$
$$\text{PRINT 999, VOL}$$
$$\text{STOP}$$
$$\text{999 FORMAT (1H0, F50.20)}$$
$$\text{END}$$

To be run on a particular computer, this program must be prefaced with a few additional statements which are written according to rules laid down by the computer center. These statements include such items as the budget number to which the computer expense is to be charged, what language the program is written in, and limits to be imposed on the time the computer should spend on the program. If you are going to use your school's computer in some of the exercises of this chapter, your professor will explain how these preface statements should be written.

Also, if you will be using any computer, chances are very good that you will be submitting your programs on punched cards. Section 8 of this chapter gives information on punching FORTRAN statements into cards. It is included in the chapter for reference purposes, so whenever you need information on card programming, you should refer to it.

EXERCISES

1. What number is printed out by the following program?

   ```
   WHLSL = 94.
   RETAIL = (WHLSL * 5.)/3.
   PRINT 999, RETAIL
   STOP
   999 FORMAT (1H0, F50.20)
   END
   ```

2. What number is printed out by the following program?

   ```
   X = 3.
   Y = 4.
   HYP = ((X * 2.) + (Y ** 2.)) ** .5
   PRINT 999, HYP
   STOP
   999 FORMAT (1H0, F50.20)
   END
   ```

3. Explain why the following program will *not* print out any number.

   ```
   AREA = A * B
   A = 2.
   B = 4.
   PRINT 999, AREA
   STOP
   999 FORMAT (1H0, F50.20)
   END
   ```

3. ASSIGNMENT STATEMENTS AND OUR FIRST PROGRAM

4. Rewrite the program in Exercise 3 so that it prints out the number 8. [*Suggestion:* Write down the statements in a different order.]

5. What number is printed out by the following program?

```
         A = 3.
         B = 4.
         C = 6.
         SIDES1 = 2. * (A * B)
         SIDES2 = 2. * (A * C)
         BOTTOM = B * C
         AREA = SIDES1 + SIDES2 + BOTTOM
         PRINT 999, AREA
         STOP
     999 FORMAT (1H0, F50.20)
         END
```

6. Rewrite the program of Exercise 5 so that it contains only one assignment statement of the form

   variable = expression

7. Write a program to compute and print out the area of a triangle whose base is 6 feet and whose altitude is 3 feet.

8. Write a program to compute and print out the area in square feet of a triangle whose base is 6 feet and whose altitude is 31 inches.

9. Write a program to compute and print out the amount of money after one year in a savings account which began at $1000 and earned 4 percent interest per year.

10. Write a program to compute and print out the amount of money after five years in a savings account which began at $1000 and earned 4 percent interest per year, compounded annually.

11. Repeat Exercise 10 for the case where the interest is compounded every three months.

12. Write a program to compute the area of a rectangle with shorter sides each 1 foot in length and perimeter 10 feet in length.

13. Write a program to compute and print out the surface area of a cylindrical tin can (including area of top and bottom) with radius of base equal to 3 inches, given that the volume of the can is 216 cubic inches.

4. Branching and Looping

The amount of time required to perform all the steps of the program in Section 3 would depend upon the speed of the computer. However, even on the slowest computers, the time required for the computations should be less than .001 second. (Printing the answer takes somewhat longer.) The tremendous speed of computers is one of the main reasons they are so useful. However, in order to take full advantage of this speed, it is necessary to have a way of directing the computer to proceed with its next task as soon as it finishes the one it is working on.

One way we could give the computer a succession of tasks would be to supply in the program of Section 3 many more pairs of values of RAD and HT, so that the computer would calculate for us many different values of VOL, all in a very short space of time. The additional values of RAD and HT are best supplied in a list following the statement END. Such a list is called *data*. This kind of use of the computer is known as *data processing*. We will not study it here because of the intricacies of getting the computer to read the data.

Another way to get the computer to continue its speedy work without pause is to make it *loop* back through the program. For example, we might be interested in calculating the volume of a cylinder of height 10 inches for various different values of the radius of the base. Suppose, for sake of argument, that we would like the volume not just for radius 3.1 but also for radius 3.2, 3.3, 3.4, $\cdots$, 4.8, 4.9, 5.0. That is, after the computer calculates the volume for radius 3.1, we want it to calculate the volume again, this time for radius 3.2, and so on. When it has made the calculation for radius 5.0, we then want it to stop.

This can be accomplished by using a rather strange-looking assignment statement

$$\text{RAD} = \text{RAD} + .1$$

to be followed by what is called a *branching* or *conditional transfer* statement. The general form of the *conditional transfer* statement is

$$\text{IF (formula) } n_1, n_2, n_3$$

The IF statement tells the computer:

(1) If the formula in parentheses is less than 0, you must do the statement numbered n_1 next.

4. BRANCHING AND LOOPING

(2) If the formula in parentheses is equal to 0, you must do the statement numbered n_2 next.

(3) If the formula in parentheses is greater than 0, you must do the statement numbered n_3 next.

Now, what do we want the computer to do? As long as RAD is less than or equal to 5.0, we want it to calculate a volume. When RAD becomes greater than 5.0, we no longer want it to calculate volumes, so we want it to STOP. The expression in parentheses is always some formula which is to be compared with 0. If, therefore, we put the formula RAD − 5.0 in the parentheses, we will be able to use the IF statement to tell the computer when to calculate and when to stop.

When RAD − 5.0 is less than 0 and when RAD − 5.0 is equal to 0, we want the computer to calculate the volume of the cylinder. That is, we want it to do the statement

VOL = (3.1416 * (RAD ** 2.)) * HT

again. To make it do that, we give this statement a number:

1 VOL = (3.1416 * (RAD ** 2.)) * HT

and then we put that same number in place of n_1 and n_2 in the IF statement:

IF (RAD − 5.0) 1, 1, n_3

On the other hand, when RAD − 5.0 is greater than 0, we want the computer to STOP, so we assign a different number, say 2, to the STOP statement and then put this number in place of n_3.

IF (RAD − 5.0) 1, 1, 2
2 STOP

Thus, we have made a total of four changes in our existing program. We have numbered two of the statements in the original program, and then following the PRINT statement we have added two more statements which cause the computer to loop back through the program nineteen times before finally coming to a STOP. The new program is shown in Figure 2, with the path of the loop indicated.

Were you at all surprised by the assignment statement RAD = RAD + .1? Many people are, because if it is viewed as an algebraic equation, it is as false as can be. Of course, we are not using it as an equation but rather as a command to the computer to do something. The statement simply tells the computer to change the value of RAD to .1 more than the present value of RAD, which is, of course, an extremely easy instruction to follow.

HT = 10.
RAD = 3.1
Statement numbered →1 VOL = (3.1416 * (RAD**2.)) * HT
PRINT 999, VOL
Statements added { RAD = RAD + .1
IF (RAD − 5.0) 1, 1, 2   } 19 times
Statement numbered →2 STOP
999 FORMAT (1H 0 , F50.20)
END

FIGURE 2. Program to compute the volume of a circular cylinder for 20 different values of the radius of the base.

EXERCISES

1. How many values of AREA will the following program print out?

    RAD = 1.
    1 AREA = 3.1416 * (RAD ** 2.)
    PRINT 999, AREA
    RAD = RAD + .1
    IF (RAD − 2.) 1, 1, 2
    2 STOP
    999 FORMAT (1H0, F50.20)
    END

2. What will the following program cause the computer to do?

    HT = 10.
    RAD = 3.1
    1 VOL = (3.1416 * (RAD ** 2.)) * HT
    PRINT 999, VOL
    RAD = RAD + .1
    IF (RAD − 5.0) 1, 2, 2
    2 STOP
    999 FORMAT (1H0, F50.20)
    END

3. What will the following program cause the computer to do?

    HT = 10.
    1 RAD = 3.1
    VOL = (3.1416 * (RAD ** 2.)) * HT
    PRINT 999, VOL

```
      RAD = RAD + .1
      IF (RAD − 5.0) 1, 1, 2
    2 STOP
  999 FORMAT (1H0, F50.20)
      END
```

4. Write a program to compute and print out the areas of triangles all of whose bases are 6 feet and whose altitudes are 1, 2, 3, 4, 5, 6, 7, 8, 9, and 10 feet.

5. Write a program to compute and print out the areas of triangles all of whose bases are 6 feet and whose altitudes are 3 feet, 3 feet 1 inch, 3 feet 2 inches, · · ·, 4 feet.

6. Write a program to compute and print out the amounts of money after 1, 2, 3, · · ·, 50 years in a savings account of $1000 that earned 4 percent interest per year, compounded annually.

7. Write a program to compute and print out the areas of rectangles all of whose perimeters are 10 feet and whose shorter sides have lengths 1, 2, 3, · · ·, 30 inches.

8. Write a program to compute and print out the surface areas of tin cans (including tops and bottoms) each having volume 216 cubic inches and having bases of radius 1, 2, 3, · · ·, 10 inches.

9. Repeat Exercise 6 for the case where the interest is compounded every three months. Have the computer print out the amount of money in the account after each three-month period.

10. Repeat Exercise 9 but have the computer print out the amount of money in the account after each year only.

11. The population of a certain town increases by 3 percent each year. Its present population is 12,354. Write a program to compute and print out the town's population for 1, 2, 3, · · ·, 50 years from the present, assuming that the growth rate remains at 3 percent over the fifty-year period.

12. Repeat Exercise 11 for 1, 2, 3, · · ·, 50 years into the past, assuming that the growth rate was 3 percent over the past fifty years.

5. An Example of a Nested Loop. Flowcharting

Sophisticated programs often contain many loops; in fact, programs are often so complicated that they have loops contained within loops.

As an example of this "nesting" of loops, let us consider a further modification of our program to one which will compute the volume of a cylinder for various values of both RAD and HT.

Suppose we would like to know the volume of the cylinder for the twenty values of RAD 3.1, 3.2, $\cdots$, 5.0 and for the ten values of HT 10, 11, $\cdots$, 19. In all, then, we would like to have 200 values of VOL computed and printed. Now the program of Figure 2 already computes twenty of these values. After those twenty values are computed, the statement

IF (RAD $-$ 5.0) 1, 1, 2

directs the computer to the statement numbered 2, which is

2 STOP

However, we no longer wish the computer to stop, but rather we want it to compute twenty *more* values of VOL, this time with HT = 11, and so on. So we erase the 2 in the IF statement and replace it with a 3. We use the number 3 to direct the computer to a new assignment statement which tells it to increase the value of HT:

3 HT = HT + 1.

Following this statement, we write down another IF statement to make the computer take a bigger loop back through the program. We want the computer to go back and do twenty more calculations for the various values of RAD, so this time we should send it back to the assignment statement RAD = 3.1, thus causing the value of RAD to be set back to the lowest value in which we are interested. We do so by giving the number 4 to that statement:

4 RAD = 3.1

and then writing down the IF statement

IF (HT $-$ 19.) 4, 4, 2

The effect of this command is to send the computer back nine times in a bigger loop. On the tenth run through this IF statement, the value of HT will be 20, and so the computer will then be directed to

2 STOP

By that time the computer will have given us precisely the $20 \times 10 = 200$ values of VOL that we wanted. Figure 3 shows the new program with the bigger loop marked on it.

5. AN EXAMPLE OF A NESTED LOOP

This example of nested loops is actually a very simple one. The nesting can get very complicated indeed. In order to portray as clearly as possible the complex maneuvers that computers are often programmed to perform, there has been developed a pictorial technique called *flowcharting*. A flowchart represents the paths and loops in a program by various shapes connected by directed line segments. A flowchart for the single-loop program of Figure 2 is given in Figure 4, and a flowchart for the double-loop program of Figure 3 is shown in Figure 5.

For programs at our level of complexity, the flowchart makes things somewhat clearer but perhaps not a great deal clearer. However, for programs containing dozens or even hundreds of statements, a well-drawn flowchart is much easier to understand than the list of statements in the program. The flowchart need not contain every detail of the program; many processes can be lumped together in a single rectangle, so that only the larger, more important paths are shown. Then, too, the words that go into the various shapes in the chart can be chosen to best help a person understand what the program does. They can be much less formal than the actual FORTRAN statements.

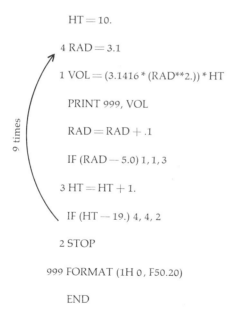

HT = 10.

4 RAD = 3.1

1 VOL = (3.1416 * (RAD**2.)) * HT

PRINT 999, VOL

RAD = RAD + .1

IF (RAD — 5.0) 1, 1, 3

3 HT = HT + 1.

IF (HT — 19.) 4, 4, 2

2 STOP

999 FORMAT (1H 0, F50.20)

END

FIGURE 3. Program to compute the volume of a circular cylinder for 20 different values of the radius of the base and 10 different values of the height.

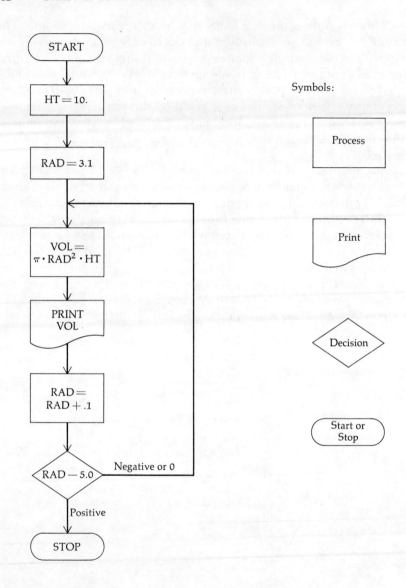

FIGURE 4. Flowchart of program in Figure 2.

5. AN EXAMPLE OF A NESTED LOOP

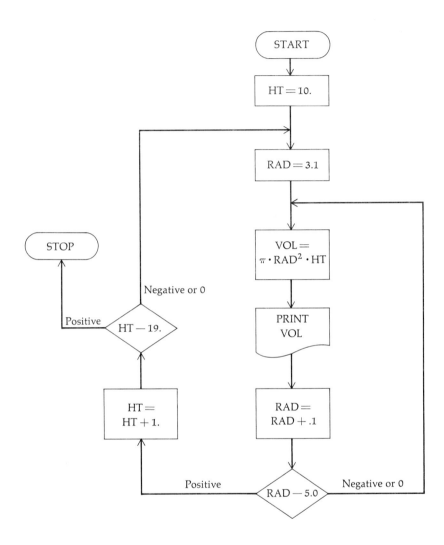

FIGURE 5. Flowchart of program in Figure 3. The two loops together cause 200 values of VOL to be computed and printed.

EXERCISES

1. What will the following program cause the computer to do?

   ```
   4 RAD = 3.1
     HT = 10.
   1 VOL = (3.1416 * (RAD ** 2.)) * HT
     PRINT 999, VOL
     RAD = RAD + .1
     IF (RAD − 5.0) 1, 1, 3
   3 HT = HT + 1.
     IF (HT − 19.) 4, 4, 2
   2 STOP
   999 FORMAT (1H0, F50.20)
     END
   ```

2. Draw a flowchart for the program in Exercise 1, showing the two loops. Explain your answer to Exercise 1 in terms of the flowchart.

3. What will the following program cause the computer to do?

   ```
     HT = 10.
   4 RAD = 3.1
   1 VOL = (3.1416 * (RAD ** 2.)) * HT
     PRINT 999, VOL
     RAD = RAD + .1
     IF (RAD − 5.0) 1, 1, 3
   3 HT = HT + 1.
     IF (HT − 19.) 1, 1, 2
   2 STOP
   999 FORMAT (1H0, F50.20)
     END
   ```

4. What will the following program cause the computer to do? Why might you want it to do such a thing?

   ```
     HT = 10.
   4 PRINT 999, HT
     RAD = 3.1
   1 VOL = (3.1416 * (RAD ** 2.)) * HT
     PRINT 999, VOL
     RAD = RAD + .1
     IF (RAD − 5.0) 1, 1, 3
   3 HT = HT + 1.
   ```

5. AN EXAMPLE OF A NESTED LOOP 265

```
    IF (HT − 19.) 4, 4, 2
  2 STOP
999 FORMAT (1H0, F50.20)
    END
```

5. The following program is intended to compute the price of 1 to 10 dozen eggs in each of five different sizes, assuming that the price per dozen of each size is 7 cents more than the price per dozen of the next smaller size. However, the program contains a number of mistakes. Find and correct all of them so that the program will do what it is intended to.

```
  4 DOZENS = 1
    PRINT 999, DOZENS
    PRPDOZ = .40
  1 PRICE = DOZENS/PRPDOZ.
    PRINT 999, PRICE
    PRPDOZ = PRPDOZ + 7.
    IF (PRPDOZ − .68) 1, 3, 3
  3 DOZENS = DOZENS + 1
    IF (DOZENS − 12) 4, 4, 2.
  2 STOP
999 FORMAT (1H0, F50.20)
    END
```

6. Write a program to compute and print out the areas of 100 triangles having bases 1, 2, · · ·, 10 feet and altitudes 1, 2, · · ·, 10 feet.

7. The field mouse population in a certain region increases at a rate of 10 percent each year up to and including the year in which the population becomes equal to or greater than 1,000,000. Once the population becomes 1,000,000 or greater, lack of food causes the population to decrease by 25 percent over each of the next two years. Then the population resumes its 10 percent growth, and so on. If the present population is 700,000, write a program to compute the field mouse population for each of the next fifty years.

8. Write a program to compute and print the amounts of money after 1, 2, 3, · · ·, 50 years in a savings account which began at $1000. and earned 4 percent interest per year for the cases where the interest is compounded annually, semiannually, and quarterly. (If you wish, print out the amounts of money after each semiannual and each quarterly period. That will make the program easier to write.)

9. In Chapter 6, Section 4, we discovered how to find the probability that at least two people from a group of n people will have the same birthday. Write a program to compute and print out this probability for the values $n = 2, 3, 4, \cdots, 70$.

6. Another Example of a Nested Loop

The example to be presented in this section is more complicated than the one of Section 5. It uses the same branching and looping techniques that we have seen in the previous two sections, so it can be omitted without your missing any new ideas. However, it is an extremely interesting example of the way loops are used in search procedures to solve problems.

We will develop the example by building up a flowchart first and then translating it into a program. The technique of drawing the chart first and then using it as a guide to write the program is common practice in computer work. A scientist who makes frequent use of the computer may not have the time to write programs, but if instead he can sketch the idea of the program in a flowchart, a professional programmer can do the work of writing the program itself.

Suppose we are interested in finding a root of the equation

$$X^3 + 2X^2 - 3X - 5 = 0.$$

The graph of this equation lies above the X-axis for $X = 10$, and it lies below the X-axis for $X = -10$. It therefore must cross the X-axis somewhere between these two values. We wish to put the computer into a loop to search out the point where the graph crosses the X-axis.

First, we will have the computer test the values of $X = 10, 9, 8, 7, 6$, and so on, until it reaches a value of X for which

$$X^3 + 2X^2 - 3X - 5 \tag{†}$$

is negative. Looking ahead, we see that that value will be 1. At that point, we will know that the root of the equation is somewhere between 1 and 2 (since the graph crosses the X-axis between those two values). To locate the root more precisely, we will then make the computer reset X to the value 2 and begin testing for the first value of the numbers $1.9, 1.8, 1.7, \cdots 1.0$ for which (†) is negative. When it finds the first such value, the root will then be located within an interval .1 units long. Then we will have the computer locate the root within an interval .01 units long, and so on until we reach the limits of accuracy of the computer or until we reach the accuracy we wish, whichever is less precise.

6. ANOTHER EXAMPLE OF A NESTED LOOP

For sake of argument, let us say we are only interested in locating the root within an interval .00001 units long, though most computers can do better. When it has performed that location (the sixth one it will be doing), we will then tell it to stop.

Now that we have the idea of the program in words, let us write down a flowchart of it, as in Figure 6. The process begins by setting

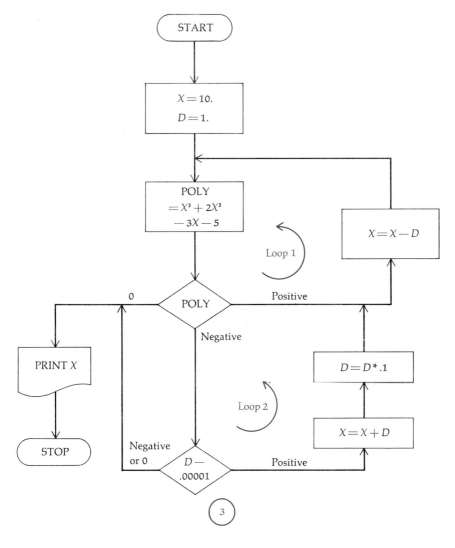

FIGURE 6. Flowchart for the process of locating a root of the equation $X^3 + 2X^2 - 3X - 5 = 0$.

the computer variable x equal to 10. At the same time we introduce another variable, D, first set equal to 1. We will use D to decrease the value of x until the variable POLY, the value of (†), becomes 0 or negative. The loop labeled 1 accomplishes this task. If and when POLY becomes negative, we send the computer into Loop 2. In this loop, we return to the previous value of x (by the statement x = x + D), make the value of D one-tenth what it was (by the statement D = D * .1), and then begin the searching process in Loop 1 again.

There are two possible ways to bring the entire process to a halt. The one way is by stumbling across a value of x for which POLY is 0. That value would be a root (to within the accuracy of the machine), so we would have the computer print it and then stop. The other way is by reaching the accuracy of .00001 which we agreed upon before. That case is signaled by D being equal to .00001, so we use the decision labeled 3 to tell the computer to print out its six-digit approximation to the root and then stop.

To write the program from the flowchart, all we need to do is imagine ourselves being carried along in the flowchart paths. First we will go along Loop 1, for which the program statements turn out to be

```
      x = 10.
      D = 1.
    1 POLY = (X ** 3.) + (2. * (X ** 2.)) - (3. * X) - 5.
      IF (POLY) , , 2
    2 X = X - D
      GO TO 1
```

Notice that after we set x = x − D, we are required to return to Statement 1. This transfer could have been accomplished by writing a "phony" decision statement like

$$\text{IF } (X) \ 1, 1, 1$$

which would send us to Statement 1 regardless of the value of x. However, this situation of an *unconditional transfer* arises so frequently that the FORTRAN language has a special statement to accomplish such a maneuver:

$$\text{GO TO } 1$$

After having spun around in Loop 1 until we reach a value of x for which POLY is no longer positive, we either go into Loop 2 or else

6. ANOTHER EXAMPLE OF A NESTED LOOP

immediately print X. Suppose we go into Loop 2. Then the program will consist of

```
    X = 10.
    D = 1.
  1 POLY = (X ** 3.) + (2. * (X ** 2.)) - (3. * X) - 5.
    IF (POLY) 3,  , 2
  2 X = X - D
    GO TO 1
  3 IF (D - .00001)  , , 4
  4 X = X + D
    D = D * .1
    GO TO 2
```

That is, in Loop 2 we set the value of X back by one notch, then make D one-tenth what it was, and then head back into Loop 1.

Now that we have followed through both loops, all we have left to do is add on the statements which cause the computer to print X and to stop. To do this, we number the PRINT statement 5 and then tie it into the two IF statements. The completed program is given below.

```
     X = 10.
     D = 1.
   1 POLY = (X ** 3.) + (2. * (X ** 2.)) - (3. * X) - 5.
     IF (POLY) 3, 5, 2
   2 X = X - D
     GO TO 1
   3 IF (D - .00001) 5, 5, 4
   4 X = X + D
     D = D * .1
     GO TO 2
   5 PRINT 999, X
     STOP
 999 FORMAT (1H0, F50.20)
     END
```

EXERCISES

1. What would the computer do if the second GO TO statement in the root-locating program should read GO TO 1?

2. Where in the root-locating program would you put a PRINT statement if you wanted the computer to print out the values of x for which (†) was evaluated?

3. Write a program to find the positive root of

$$X^2 - 2 = 0$$

to five decimal places accuracy. What is this root usually called? Is there another way you can use the computer to find this number?

4. Why can you not write a program to find a root of

$$X^2 + 100 = 0?$$

5. Write a program to find, to five decimal places accuracy, the value of X which makes

$$35 - 2X^2 + 9X \qquad (††)$$

as large as possible. [*Hint:* Evaluate the expression for $X = 100$, 99, 98, $\cdots$ until you finally arrive at a value of X which causes (††) to *decrease* from its value for the previous X. Then reset X by *two* notches, and go through a loop similar to Loop 2 in the text.]

6. Tell why it is not sufficient in Exercise 5 to reset X one notch only.

7. Write a program to locate the maximum area which can be enclosed by a rectangle of perimeter 10 yards. [*Hint:* Program the computer to search for the length of a side which maximizes the area.]

8. Modify the program in Exercise 7 of Section 5 so that the computer prints out at the very end the largest field mouse population over the fifty-year span.

9. Write a program to locate the minimum surface area of a tin can (including top and bottom) which has 216 cubic inches capacity. [*Hint:* Program the computer to search for the radius of the base which minimizes the area.]

10. Problems of the kind stated in Exercise 9 are very important in cutting production costs (and amounts of waste). To be more practical, however, one should allow for the possibility that the top and bottom of the can are made from material with a cost per square inch different from that of the side. Assuming the cost

of top and bottom per square inch is three times that of the side, repeat Exercise 9 to find the radius of the base which minimizes the cost of the tin can.

7. Computers—Past, Present, and Future

One of the earliest devices used in computing was the *abacus*. It was used by the Greeks before the time of Christ, by Europeans of all nationalities until the fifteenth century, and by various peoples even today. The principle of the abacus, with its beads slid on wires, is the same principle employed in adding machines, desk calculators, and computers. In a sense, then, the abacus can be regarded as an ancestor of our present computers.

In the seventeenth century, the principle of the abacus was incorporated into mechanical calculating devices. One such device was an adding machine made by Blaise Pascal (1623–1662), and the other was a machine which could perform addition, subtraction, multiplication, and division, made by Gottfried Wilhelm Leibniz (1646–1716). Leibniz's machine was the forerunner of our present-day electromechanical desk calculators.

However, as we have seen in this chapter, there is an important difference between a desk calculator and a computer. The desk calculator requires a human operator to perform each successive arithmetic operation on it, while a computer is capable of performing a whole string of arithmetic operations, making decisions every step of the way as to whether it should continue to calculate and what operation it should next perform.

The first person to envision a calculator performing such a sequence of operations automatically was the English mathematician Charles Babbage (1792–1871). Babbage designed but did not succeed in building a purely mechanical device which was essentially the same as a modern computer. He called his attempted machine the "analytical engine." His lack of success in building the machine was due to the fact that it was not possible to mill the parts to the close tolerances needed to make the machine work. Babbage's idea could not be realized until the invention of suitable electric circuitry a century later.

The first electrically operated computer was the Mark I, at Harvard University, installed in 1944. It was soon surpassed by the ENIAC (Electronic Numerical Integrator And Calculator) at the University of Pennsylvania, installed in 1945. The ENIAC was the first computer to use vacuum tubes. (Mark I used relays.) Computers using vacuum tubes are known as *first-generation* computers. Transistorized computers, known as *second-generation* computers, appeared in 1958. They were a great step forward in speed, reliability, and compactness. *Third-generation* computers, which use integrated circuits (circuits as complicated as transistor radios but entirely contained within a piece of silicon one-tenth of an inch square), were introduced in 1965. Currently, a fourth generation of computers is being developed, to appear perhaps in the mid-1970s.

Simultaneously with the rapid physical development of computers (or "hardware"), there has been a rapid development of programming techniques (or "software"). The greatest of all programming strides was the development of the first high-level language, FORTRAN. As we mentioned, there are now two major high-level languages, ALGOL and PL/I, which are contenders with FORTRAN for popularity. ALGOL (Algorithmic Language), developed around 1960, has been designated the official international scientific programming language. PL/I (Programming Language I), developed around 1968, is a universal language suited for both scientific and business (financial) purposes. So far, PL/I is usable only on IBM System/360 computers.

Besides these languages, there are some languages that are almost conversational. BASIC (Beginner's All-Purpose Symbolic Instruction Code) is the best-known of these. There are a number of computer systems on which it cannot yet be used, and there are a few types of problems for which it is not very suited. However, the experience of writing programs in BASIC is a joy. If you expect to become an occasional user of a computer, and if your computer center has BASIC as one of its languages, you might consider making it your first programming language, rather than continuing to learn FORTRAN. (Most of what we have covered in this chapter is applicable to BASIC with only slight changes.)

One popular application of computers that has not been illustrated in this chapter is *simulation*. Various complicated phenomena such as card games, business enterprises, the stock market, and wars can be studied by programming the computer according to the rules of the game, the conflict, or whatever. Whenever an event of uncertain outcome appears, the computer determines the outcome by picking a number from a random number generator (the electronic version of a lottery) and then proceeds to the next event. The game or war or whatever is "played" again and again by the computer, and the various final outcomes are tabulated and later studied. A classic example of a simulation was the one by Edward Thorp, who simulated the card game Blackjack and discovered from his results a "system" for winning. (See his book, *Beat the Dealer*, for details.) His system worked so well that the Nevada casinos were forced to alter the rules of the game.

I am reminded of a story about a simulation which perhaps illustrates what it is about computers that really fascinates the layman. Once upon a time, a company contracted with a branch of the armed services to do a simulation of a "conflict." Once the program was running, the company gave a demonstration of a computer run to some visiting representatives of the particular armed service. The computer did its "thing," and at the end of the run a few pages of numbers were printed out. And that was that. For some reason, the visitors did not seem impressed. So, before the next batch of visitors was due, the programmers added instructions at the end of the program to rewind tapes, make panel lights flash, and so on. The next demonstration ended in a glorious display of motion and color, and the visitors actually volunteered how impressed they were.

In any extended discussion of computers, there will arise the question, "Do computers think?" This question can be answered by either a yes or a no, depending upon one's definition of the word "think." We have already seen that computers are capable of making simple decisions in the FORTRAN IF statement. Therefore, if making simple decisions is thinking, then computers think. On the other hand, for those who have a vested interest in maintaining that only men are capable of thinking, a slightly more restrictive definition of the word think is all that is required in order to say that computers do not think. Of course, as computer technology develops, the definition of thinking will have to be made more and more restrictive to keep computers from having thoughts. Recently computers have been programmed to discover proofs of certain (somewhat elementary) theorems.

In just twenty-five years, the number of computer users has grown from a few dozen pioneers to thousands upon thousands of people from all walks of life. The number of computers and the speed of processing have kept up somewhat with the increased patronage, but the typical user can expect to have several hours to a full day wait while his program is being run.

Happily, the technology which made it possible to build the large computers has also made it possible to build memories and branching and looping capabilities into machines that are not much larger than the old desk calculators. These machines are generally called *programmable calculators*. They can be programmed to perform much the same kinds of steps large computers can perform—just not as many. Some of them have sufficient program capacity to handle any program we have discussed in this chapter.

If you wish to build on what you have learned in this chapter and if you have access to such a calculator, you can gain much valuable experience on it in a short time. The techniques of programming a calculator use—in some way or other—the same keys that are used to do arithmetic on the machine. The techniques are very easy to learn by reading the manual for the particular machine, but since they differ from one machine to the next, we will not attempt to discuss them here.

Another method for eliminating the waiting time in running programs is the use of *time-sharing terminals*. Under the time-sharing system, a single computer is connected to many typewriter terminals. Computer users sit at the typewriters and write out their programs. The computer divides its time between those terminals from which programs are being submitted. It seldom happens that a user will receive immediate or undivided attention from the computer, but because of the computer's speed, it *appears* to the user that the computer is instantly responding to his program statements. Time-sharing programs are usually written in a conversational language such as BASIC, so as to make life even easier for the user. Programming in time-sharing BASIC is much like having a conversation with the computer. If the user submits an incorrect statement to the computer, back comes a typed message telling exactly what *part* of the statement is incorrect.

8. Card Programming

A very common way of submitting a program to a computer is to use a deck of "IBM" cards, with one statement punched into each card. The advantages of this method are fairly clear. The card deck is a convenient means for storing the program and for carrying it to the computer center whenever it is to be run. Also, in the processes of correcting mistakes *(debugging)* and modifying the program, it is very easy to pull out cards containing unwanted statements and insert other cards containing new statements.

A typical FORTRAN card, punched for Statement 1 of Figure 2, is shown in Figure 7. The columns marked 2 through 5 are used for the statement number if the statement has been given a number. The statement number must always be right-justified; that is, its units digit must be in Column 5. The columns numbered 7 through 72 are used for the FORTRAN statement. In the rare case that you cannot fit a FORTRAN statement into the $72 - 6 = 66$ spaces allowed on the card, you can punch the number 2 in column 6 of a second card and then continue the FORTRAN statement onto that card, again in the columns 7 through 72.

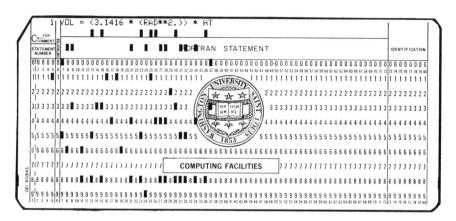

FIGURE 7. Punched card containing Statement 1 of the program in Figure 2. An electrically operated card reader senses which holes have been punched in the card and then relays the information to the computer. The statement printed on the top of the card is for the information of the programmer.

The computer pays no attention to blanks in a FORTRAN statement. You can leave as few or as many blanks as you want, depending on whether you are economizing on space or are trying to make the printed line at the top of the card readable. Since blank spaces in FORTRAN statements mean nothing to a computer, it is all-important that those separator commas used in various statements not be left out. They are needed to tell the computer where one number ends and the next begins. Since commas are used to separate numbers, they can never be used within one number. For example, *ten thousand* must be written 10000. or (if you care to use a blank space as a separator) 10 000.

While we are on the subject of punctuation, remember also that the period represents a decimal point only and not the end of a statement or anything else. Putting a period at the end of a statement out of habit will completely befuddle the computer unless, of course, the period serves as a decimal point. On the other hand, every constant must have a decimal point, even if the constant is a whole number.

To make the card punching operation go smoothly, it is a good idea first to write out your program on sheets of special FORTRAN coding paper. A portion of a sheet of this paper is shown in Figure 8, with the program from Figure 2 written on it. In the serious use of computers, the programmer seldom uses the card punching machine. He normally writes his program on the coding sheets, and from these a keypunch operator produces the card deck. This division of labor makes it especially necessary that the programmer print legibly when he writes his program on the coding sheets.

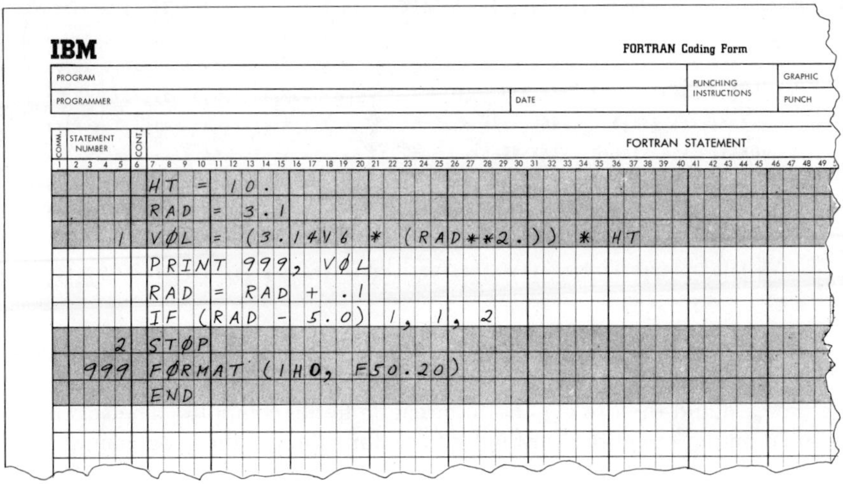

IBM FORTRAN Coding Form

PROGRAM | PUNCHING INSTRUCTIONS | GRAPHIC
PROGRAMMER | DATE | PUNCH

COMM. / STATEMENT NUMBER	CONT.	FORTRAN STATEMENT
		HT = 10.
		RAD = 3.1
1		VØL = (3.1416 * (RAD**2.)) * HT
		PRINT 999, VØL
		RAD = RAD + .1
		IF (RAD - 5.0) 1, 1, 2
2		STØP
999		FØRMAT (1H0, F50.20)
		END

FIGURE 8. Portion of a FORTRAN coding form. The statements on the form are those of Figure 2. The statements are written out line by line, with the characters placed in the same spaces the programmer wishes them to be in on the punched cards.

CHAPTER THIRTEEN

An Introduction to Calculus

1. Introduction

The single most important mathematical subject taught in college is calculus. No other branch of mathematics has had so profound an effect on the civilization of this planet. All of the technology that has become such an important part of life today was made possible by the invention of calculus. If a person wishes to understand the origins of our technological culture, it is necessary that he have at least a passing acquaintance with the branch of mathematics which contributed so heavily to its development.

The primary purpose of this chapter is to give you that passing acquaintance—to give you the general flavor of calculus. If you find that you would like to go further into calculus after this course, it will also have given you a little preparation for the course. Calculus courses usually go very fast and cover a tremendous amount of material. The students in them are mainly highly motivated science and engineering majors who are willing to spend long hours learning the methods for doing the 2000 exercises in their 600 page textbook. In other words, the standards are high. Given the situation, it can be a help to have had some orientation beforehand.

278   CHAP. 13   AN INTRODUCTION TO CALCULUS

2. Numbers, Functions, and Graphs

Calculus is a heavily number-oriented branch of mathematics. As we begin our brief study of calculus, let us reach an agreement about what we will be calling a *number*. For us, a number will be any one of those creatures-of-arithmetic that can be written down in the decimal notation system. Some examples of numbers are

$$1 \quad 12 \quad -5 \quad -6.6666\cdots \quad .217217217\cdots$$
$$1.14472988584940017414342735135305871164 72\cdots.$$

Notice that the digits to the right of the decimal point can continue indefinitely (as in $-6.6666\cdots$); furthermore, they can continue indefinitely without even giving a hint of a regular pattern (as in $1.14472\cdots$).

Those of you who have had a high school mathematics program oriented toward "structures" may have heard the collection of all these numbers referred to as the *system of real numbers*. We will refer to them simply as *numbers*.

The basic mathematical object with which calculus is concerned is the *function*.

DEFINITION. Consider a collection D of numbers. A *function f on D* is a rule which assigns to each number x belonging to D a number $f(x)$. The collection R made up of all the numbers $f(x)$ is called the *range* of f. The collection D is called the *domain* of f.

It is possible to generalize the definition of a function up to the sky and down into the ground, but the definition we have given here is sufficient for beginning purposes. Let us look at some examples to fix the idea of *function* in our minds. The definition may be new, but we have known examples which satisfy the definition since early in high school.

Example 1

$$f(x) = x^3.$$

To every number x, this function assigns the number x^3. For instance

$$f(2) = 8 \quad f(0) = 0 \quad f(-\tfrac{1}{2}) = -\tfrac{1}{8}.$$

This function has for its domain the collection of all numbers. Its range, as well, consists of the collection of all numbers.

Example 2

$$f(x) = 16x^2.$$

To every number x, this function assigns the number $16x^2$. For instance,

$$f(\tfrac{1}{2}) = 4 \qquad f(-1) = 16 \qquad f(0) = 0.$$

This function has for its domain the collection of all numbers, but its range is restricted to those numbers which are 0 or greater.

Example 3

$$f(x) = x^2 - x + 1.$$

To every number x, this function assigns the number $x^2 - x + 1$. For instance

$$f(0) = 1 \qquad f(\tfrac{1}{3}) = \tfrac{7}{9} \qquad f(-1) = 3.$$

This function has for its domain the collection of all numbers. At present, we do not have the means to determine exactly what its range is. It turns out that not all numbers are in its range; in Section 4 of this chapter we will develop a tool which will enable us to prove this and in fact to determine precisely what the range of f is.

Example 4

$$f(x) = \sqrt{x}.$$

To every number x which is greater than or equal to 0, this function assigns the number y such that $y^2 = x$ and such that y is greater than or equal to 0. (This function is sometimes called the "positive" square root function.) For instance

$$f(0) = 0 \qquad f(\tfrac{1}{4}) = \tfrac{1}{2} \qquad f(-1) \text{ is not defined}$$
$$f(2) = \text{approximately } 1.414.$$

This function has for its domain the collection of all numbers that are 0 or greater. Its range also consists of all numbers that are 0 or greater.

There are more complicated examples of rules satisfying our definition of function, but these will be enough for present purposes. (We will have a special section later devoted to two other, very important examples of functions.) Example 1 shows that it is possible for both the domain and the range of a function to consist of the col-

280   CHAP. 13   AN INTRODUCTION TO CALCULUS

lection of all numbers, while Example 4 shows that it is possible that neither the domain nor the range of a function will consist of all numbers.

There are many ways to draw pictures of functions. One picture that I and some others like to draw on the blackboard to get across the idea of the definition is shown in Figure 1. A function f is a device that accepts certain numbers x and turns them into other numbers $f(x)$. (Sometimes, of course, the same number will pop out as was put in.)

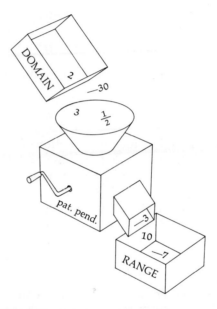

FIGURE 1. A function viewed as a device for converting numbers into other numbers. The numbers falling out of the machine, $-3, 10, -7$, might be $f(0), f(-\frac{1}{2})$, and $f(-5)$, respectively.

A very useful way of portraying a particular function is the familiar graphing technique invented by René Descartes. (We see so many graphs today from our earliest years that it is hard to realize what a brilliant, original idea it was that Descartes had.) To draw the graph of a function, first draw two lines, one horizontal and one vertical, calling the point at which they cross the *origin*. Lay out a uniform scale on both lines, with the number 0 at the origin and the positive numbers going to the right and going upward, as in Figure 2. Each line, together with its scale, is called an *axis*.

Then, for each x in the domain of f, mark the point which is above (or below) the number x on the horizontal axis and which is

2. NUMBERS, FUNCTIONS, AND GRAPHS   281

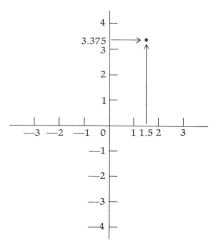

FIGURE 2. Illustration of the two axes, together with one point which lies on the graph of $f(x) = x^3$.

level with the number $f(x)$ on the vertical axis. For instance, in drawing the graph of the function x^3 we would mark a point above the number 1.5 on the horizontal axis and level with the number $1.5^3 = 3.375$ on the vertical scale. Figures 3, 4, and 5 show the graphs (or, strictly speaking, portions of the graphs) drawn for the functions

$$x^3, \quad x^2 - x + 1, \quad \sqrt{x},$$

respectively.

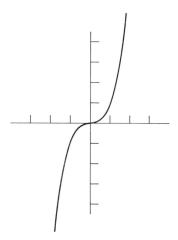

FIGURE 3. Portion of the graph of $f(x) = x^3$.

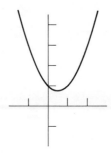

FIGURE 4. Portion of the graph of $f(x) = x^2 - x + 1$.

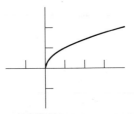

FIGURE 5. Portion of the graph of $f(x) = \sqrt{x}$.

Notice that each graph is a remarkably smooth, regular curve. Most functions encountered in practical work have graphs with a smooth, flowing appearance. To draw such a graph, all one needs to do is mark a few select points and then "connect the dots" in a smooth, flowing manner. (There is a special flexible ruler made for draftsmen who do many curve and function drawings. You just bend the gadget until it follows the dots and then draw along it.)

Figure 5, however, shows one exception to the smooth-and-flowing rule of thumb. The graph ends abruptly at the origin. Why could it not continue, as in Figure 6? The answer is that Figure 6 is *not* the

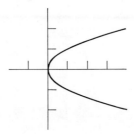

FIGURE 6. The graph in Figure 5 can be continued, but the resulting curve is no longer the graph of a function.

2. NUMBERS, FUNCTIONS, AND GRAPHS

graph of a function. For example, corresponding to the value $x = 4$, there are two points marked in Figure 6, one at level 2 and the other at level -2. The definition of a function, however, allows $f(4)$ to be only one number, and that is the reason the graph in Figure 5 stops at the origin.

EXERCISES

1. Draw (reasonable portions of) the graphs of the following functions:
   (a) $f(x) = x^2 - 2x + 1$.
   (b) $f(x) = x^2$.
   (c) $f(x) = 2x + 3$.
   (d) $f(x) = \dfrac{1}{x}$.

2. Which of the following curves are graphs of functions? For those which are not graphs of functions, tell why they are not.

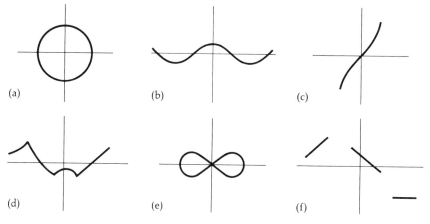

3. Let the following figure be the (complete) graph of a function. Trace this graph onto a sheet of paper and draw colored line segments on the horizontal and vertical axes to indicate the domain and range of the function.

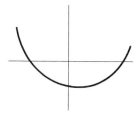

4. Repeat Exercise 3 for each of the curves in Exercise 2 which are graphs of functions.

5. Suppose I have 30 feet of wire fence with which I wish to make a rectangular pen. If two opposite sides of the pen are each x units long, how long will the other two sides be? Express the area of the pen as $f(x)$. What is the domain of this function? [*Note:* The domain does *not* consist of all real numbers.]

6. Graph the function of Exercise 5. Draw colored line segments on the horizontal and vertical axes to indicate the function's domain and range.

7. A certain state wishes to erect a new office building, which will cost $200,000 for the first floor and $115,000 for each additional floor. Suppose x dollars are appropriated by the legislature for constructing the building, and that the maximum number $f(x)$ of floors for that amount of money are built. Graph $f(x)$ as a function for values of x less than $1,000,000.

8. The function in Exercise 7 differs in one important way from the one in Exercise 5 in having *jump discontinuities* (vertical breaks in the graph). How many jump discontinuities does the function in Exercise 7 have for x less than $1,000,000?

9. In Exercise 7, suppose the excess money (that which cannot go into the building of one last floor) is used for landscaping. Graph the amount of money $f(x)$ spent on landscaping for all values of x less than $1,000,000, where x is the money appropriated for the building.

10. According to Newton's law of gravitation, the force of attraction $f(x)$ exerted on a one pound mass by the earth is approximately

$$\frac{16{,}000{,}000}{x^2},$$

where x is the distance of the mass in miles to the center of the earth. Graph this function for all values of x between 4000 and 25,000 miles. What force is exerted on a 1-pound mass located 8000 miles from the center of the earth? Answer this same question for 1-pound masses located 4000 and 16,000 miles from the center of the earth. By the way, what *is* the radius of the earth on which we live?

3. Rate of Change

The function

$$f(x) = 16x^2$$

is an extremely important one in the history of science. Galileo Galilei (1564–1643) discovered that this function describes (to a good approximation) the motion of free-falling objects. This discovery marks the *beginning* of the mathematical study of motion. (If you do not know how motion was studied up until Galileo's time, you might be interested in reading some of Aristotle's *Physics*, which was accepted almost without question by the intellectuals of that time. No slur upon Aristotle is intended; he could not possibly have been correct in everything he wrote. The number of times he was right is staggering.)

By means of ingenious experiments of his own invention, Galileo found that if air resistance is not great, an object dropped without throwing will fall a distance of $16x^2$ feet in x seconds. Table 1 lists the various distances an object will have traveled in 1 second, 2 seconds, 3 seconds, and 4 seconds after it is dropped. Table 1 also lists the distance covered in *each* second by the object. Notice that the distance covered in each second increases as the seconds tick by. That is, the object is picking up speed as it falls.

TABLE 1. Total distance covered at the end of each second and distance covered within each second.

Seconds Elapsed	Distance Covered	Second	Distance Covered
0	0	1st	16 − 0 = 16
1	16	2nd	64 − 16 = 48
2	64	3rd	144 − 64 = 80
3	144	4th	256 − 144 = 112
4	256		

Simple as this last observation is, it opens the door to calculus; but it does so by posing a tough question that we must answer. Speed is usually defined as distance covered divided by time elapsed, as is indicated by the units in which it is measured: miles per hour, feet per second, furlongs per fortnight, and others. Now it is a simple matter to divide 80 feet by 1 second and conclude that during the 3rd second the object fell at a speed of 80 feet per second. But does this number, 80 feet per second, tell us what we really want to know? Remember, all through the 3rd second, the object was speeding up. The number 80 feet per second is an *average* speed. It is *not* the speed of the object when $x = 2$ seconds, and it is *not* the speed of the object when $x = 3$ seconds. Presumably, at time $x = 2$ the object was falling slower than 80 feet per second, and at time $x = 3$ the object was falling faster than 80 feet per second. That way the speeds could average out to 80 feet per second over the 3rd second.

But now we have a logical difficulty. You see how much we would like to talk about the speed at $x = 2$ seconds. In fact, we have been talking about speed-at-a-particular-time as though we already know what that means. But if we pause and think deeply for a while, we discover that we do not know what speed-at-a-particular-time means. Remember, speed is normally defined as

$$\text{speed} = \frac{\text{distance traveled}}{\text{time elapsed}}.$$

If we try letting 0 seconds elapse and thus travel 0 feet, this definition of speed leads to

$$\text{speed} = \frac{0}{0}.$$

But what is the quotient 0/0? Is it 3? ½? −10? $\sqrt{2}$? It is impossible to assign a unique value a to the quotient 0/0, because saying

$$a = \frac{0}{0}$$

is the same as saying that a satisfies

$$a \cdot 0 = 0. \tag{†}$$

But every number a satisfies (†).

Therefore, it appears that we will need to find a new definition for speed-at-a-particular-time, because the old definition of speed only gives speed over an interval of time. We should, of course, make use of the old definition of speed in our definition of speed-at-a-particular-time, because we want the two kinds of speeds to bear some family resemblance to each other. Perhaps if we apply the old definition of speed to a very small elapsed time, we just might get something.

For example, if we take the average speed over 1/10 second

$$\frac{16(2.1)^2 - 16 \cdot 2^2}{2.1 - 2},$$

perhaps during that short amount of time the speed we are hoping to define (and calculate) will not grow very much. In other words, the average speed over 1/10 second should be fairly close to what we hope to define as the speed at $x = 2$ seconds. What does the division in the above fraction produce?

$$\frac{16(2.1)^2 - 16 \cdot 2^2}{2.1 - 2} = \frac{70.56 - 64}{.1}$$
$$= 65.6 \text{ feet per second.}$$

So far so good. But we should be able to do even better by letting the elapsed time be only 1/100 second:

$$\frac{16(2.01)^2 - 16 \cdot 2^2}{2.01 - 2} = 64.16 \text{ feet per second.}$$

Now try 1/1000 second:

$$\frac{16(2.001)^2 - 16 \cdot 2^2}{2.001 - 2} = 64.016 \text{ feet per second.}$$

Try 1/10,000 second:

$$\frac{16(2.0001)^2 - 16 \cdot 2^2}{2.0001 - 2} = 64.0016 \text{ feet per second.}$$

Can you guess what will happen with 1/100,000 second, and so on? It certainly looks as though by making the elapsed time very small, we can obtain an average speed for that elapsed time as close as we please to the number 64. Although we still have not made a formal definition of speed-at-a-particular-time, it looks as though whatever definition we make had better give the value 64 as the answer when $f(x) = 16x^2$ and $x = 2$.

To ensure this, let us try to put together a definition that makes use of the process we went through to arrive at the number 64.

DEFINITION. By the *derivative* of the function f at the value x, we will mean that number which

$$\frac{f(x') - f(x)}{x' - x} \qquad (1)$$

approaches as $x' - x$ approaches 0. We will denote the derivative of f at the value x by $f'(x)$.

DEFINITION. If $f(x)$ represents the position of an object as a function of time x, we will also call $f'(x)$ the *speed of the object at time x*.

Of course, it may very well happen that the quotient (1) will not approach any number as $x' - x$ approaches 0. Each individual function

must be checked to see what happens. Let us first check the function
$$f(x) = 16x^2$$
completely. We have
$$\frac{f(x') - f(x)}{x' - x} = \frac{16x'^2 - 16x^2}{x' - x}$$
$$= 16\frac{x'^2 - x^2}{x' - x}$$
$$= 16\frac{(x' - x)(x' + x)}{x' - x}$$
$$= 16(x' + x).$$

Now as $x' - x$ approaches 0, x' must approach x, so $16(x' + x)$ must approach
$$16 \cdot 2x = 32x.$$
Therefore, the speed at x seconds is
$$f'(x) = 32x$$
for *all* values of x greater than or equal to 0.

It is interesting to see how nicely this speed at time x fits in with the average speeds of the falling object. Table 2 is the same as Table 1, except for the column added to give the speed at 0, 1, 2, 3, and 4 seconds. In each case, the speed during an interval is one-half the sum of the speeds at the beginning of the interval and at the end of the interval. That is, the average speed during the interval is exactly half-way between the speeds at the beginning and the end of the interval.

TABLE 2. Comparison of the speeds-at-particular-times with the average speed over each second.

x Seconds Elapsed	Distance Covered	Speed at x Seconds	Second	Average Speed over Second
0	0	0	1st	16
1	16	32	2nd	48
2	64	64	3rd	80
3	144	96	4th	112
4	256	128		

Galileo knew about the speed-at-a-particular-time of a falling object, so he had at least a glimmer of the idea of the derivative. However, it was Isaac Newton (1642–1727) who first hit upon the definition of the derivative in general. The year of his discovery of the *fluxion* (his name for the derivative) was 1671. Less than five years after Newton's discovery of the derivative, Gottfried Wilhelm Leibniz independently made the same discovery. The remarkable coincidence in time of these two flashes of inspiration indicates that previous discoveries in mathematics and physics must have brought science to a level that made the discovery of calculus possible. In line with this observation, it should be mentioned that several mathematicians —Pierre de Fermat, John Wallis (1616–1703), and Isaac Barrow (1630–1677)—came very close to discovering the derivative and the other ideas of calculus, all within the thirty years prior to Newton's work.

Newton and Leibniz not only discovered the derivative, but they also developed practical methods for computing derivatives. You may have noticed that our computation of the derivative of $16x^2$ involved a step in which we factored $x'^2 - x^2$ into the product

$$(x' - x)(x' + x).$$

Had we not thought of factoring, we would have been stopped at that point. One of the methods of computing derivatives helps avoid problems of that sort.

The idea is to introduce another number

$$a = x' - x.$$

Since

$$x' = x + a,$$

the fraction (1) can be rewritten as

$$\frac{f(x+a) - f(x)}{a}. \tag{2}$$

Calculating what (2) approaches as a approaches 0 is usually very mechanical and very easy.

For example, for $f(x) = 16x^2$, we have

$$\frac{f(x+a) - f(x)}{a} = 16 \frac{(x+a)^2 - x^2}{a}$$

$$= 16 \frac{x^2 + 2ax + a^2 - x^2}{a}$$

$$= 16 \frac{2ax + a^2}{a}$$

$$= 16 (2x + a)$$

which approaches $32x$ as a approaches 0.

Let us try the method on $f(x) = x^3$.

$$\frac{f(x+a) - f(x)}{a} = \frac{(x+a)^3 - x^3}{a}$$

$$= \frac{x^3 + 3x^2a + 3xa^2 + a^3 - x^3}{a}$$

$$= \frac{3x^2a + 3xa^2 + a^3}{a}$$

$$= 3x^2 + 3xa + a^2$$

which approaches $3x^2$ as a approaches 0.

Even a sum of several powers presents no difficulty. Let us demonstrate by finding the derivative of $f(x) = x^2 - x + 1$.

$$\frac{f(x+a) - f(x)}{a} = \frac{(x+a)^2 - (x+a) + 1 - (x^2 - x + 1)}{a}$$

$$= \frac{x^2 + 2xa + a^2 - x - a + 1 - x^2 + x - 1}{a}$$

$$= \frac{2xa + a^2 - a}{a}$$

$$= 2x + a - 1$$

which approaches $2x - 1$ as a approaches 0.

EXERCISES

1. Find the derivatives of the functions

   $f(x) = x^2 - 1$ $\qquad$ $f(x) = x^3 + x$
   $f(x) = 2$ $\qquad$ $f(x) = x$.

2. Find the derivatives of the functions

   $f(x) = x^2$ $\qquad$ $f(x) = x^4$ $\qquad$ $f(x) = x^5$.

   On the basis of your work, plus the example $f(x) = x^3$ in the text, take a guess at what the derivative of the function

   $$f(x) = x^n$$

   is for any natural number n.

3. Find the derivatives of the functions

   $f(x) = x + 1$ $\qquad$ $f(x) = 2x + 2$ $\qquad$ $f(x) = 3x + 3$
   $f(x) = 2x^2$ $\qquad$ $f(x) = 3x^2$ $\qquad$ $f(x) = 5x^2$.

   On the basis of your work, try to formulate a rule which would be helpful in the process of calculating derivatives.

4. Find the derivatives of the functions

$$f(x) = x^2 + x \qquad f(x) = x^3 + x^2 \qquad f(x) = x^3 + x.$$

On the basis of your work, try to formulate a rule which would be helpful in the process of calculating derivatives.

5. What is the derivative of the speed, $32x$, of a falling object? Say in words what your result means.

6. A spherical hailstone or raindrop grows in such a fashion that its radius is proportional to the time elapsed since it began growing. That is, if x is the time in hours and r is the radius in inches, we might have

$$r = \frac{1}{100}x.$$

Write the volume of the hailstone or raindrop as a function of the time x. Find the derivative of the volume function. Say in words what your result means.

7. Find the derivative of

$$f(x) = \tfrac{1}{3}x^5 + 3x^3 + x^2 - 1001$$

by applying the rules you guessed in Exercises 2, 3, and 4.

8. Compute the derivative of $f(x)$ in Exercise 7 by the tried-and-true method using (2), to see that you obtain the same answer as in Exercise 7.

9. Give a proof of the formula for $f'(x)$ you guessed in Exercise 2.

10. A ladder 13 feet long was standing vertically against a wall. However, a dog whose leash is tied to the foot of the ladder is pulling the foot of the ladder away from the wall at a speed of 2 feet per second. The result is that the top of the ladder is sliding down the wall. When the foot of the ladder is 5 feet from the wall, how fast is the top of the ladder moving downward? [*Hint:* The function describing the height involves a square root. If you find you cannot discover what the derivative of that function is, first try taking the derivative of $f(x) = \sqrt{x}$ to learn how to find the derivative of a square root function.]

11. (For those familiar with computer programming.) Write a FORTRAN program to calculate the quotient (2) for

$$f(x) = \tfrac{1}{3}x^5 + 3x^3 + x^2 - 1001;$$
$$x = 1, 2, \text{ and } 3 \text{ and } a = .01, .001, .0001, \text{ and } .00001.$$

(That is, you will be calculating values of (2) a total of 12 times.)

12. (For those with access to a computer.) Run the program of Exercise 11, and compare the printout with $f'(1)$, $f'(2)$, and $f'(3)$.

4. The Slope of a Tangent Line. Maximum and Minimum

The derivative $f'(x)$ can be given a very nice interpretation in terms of the graph of f. In Figure 7 we have located two points on the graph of a function f. Through these two points we have drawn a line L.

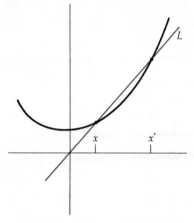

FIGURE 7. A line through two points on the graph of a function.

Now, the *slope* of any nonvertical line is defined to be the ratio

$$\frac{\text{rise}}{\text{run}},$$

where "run" is the horizontal distance between two points on the line, and "rise" is the vertical distance between those same two points, as is illustrated in Figure 8.

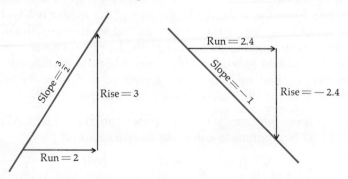

FIGURE 8. Two lines, one of positive slope and one of negative slope.

4. THE SLOPE OF A TANGENT LINE

By definition, then, the slope of the line L in Figure 7 is the ratio

$$\frac{f(x') - f(x)}{x' - x}$$

which we recognize as the fraction (1) in our definition of the derivative. What happens as we let x' approach x? The line through the two points on the graph begins to turn, getting closer and closer to the line T which is tangent to (or "touching") the graph at the point above the number x. As the line L turns, its slope must therefore approach the slope of the line T. Since the derivative was *defined* to be the number approached by the ratio

$$\frac{f(x') - f(x)}{x' - x},$$

it follows that the derivative $f'(x)$ is the slope of the tangent line T. See Figure 9.

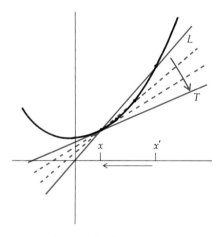

FIGURE 9. As x' approaches x, the line L swings around to approach the line T. (This happens also if x' is to the left of x.)

This geometric interpretation of $f'(x)$ is very useful, because it gives us the idea of a method by which we can find the maximum and minimum values of a function f (if these values exist). Figure 10 illustrates two ways in which a function can achieve a maximum value. In Figure 10(a) the graph has a rounded top, and the line tangent to the graph at the highest point is perfectly horizontal. Since the derivative is the slope of the tangent line, the derivative $f'(x)$ for that highest point of the graph must be equal to 0.

In Figure 10(b), the graph has a pointed top, and so it is impossi-

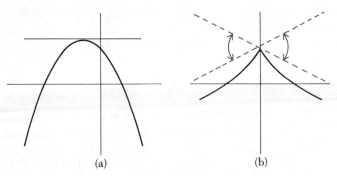

FIGURE 10. Two ways in which a function can attain a maximum value. (a) Derivative = 0 at maximum. (The function graphed is $f(x) = 1 - x - x^2$.) (b) Derivative does not exist at maximum. (The function graphed is $f(x) = 1 - \sqrt[3]{x^2}$.)

ble to draw a unique tangent line at that point, because a line balanced up there would be free to "wobble." In terms of the ratio

$$\frac{f(x') - f(x)}{x' - x}, \tag{1}$$

the wobbling means that the fraction (1) cannot approach a number as x' approaches x. That is, the derivative $f'(x)$ does not exist at the highest point of the graph in Figure 10(b).

It can be proved that if f has its maximum value at x, then one of the following is true:

(1) $f'(x) = 0$, as in Figure 10(a).
(2) $f'(x)$ does not exist, as in Figure 10(b).
(3) The domain of f either begins or ends (at least temporarily) at the value x. This third case is illustrated in Figure 11 for the function $f(x) = -(\sqrt{x})^2$.

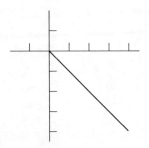

FIGURE 11. Graph of the function $f(x) = -(\sqrt{x})^2$, illustrating occurrence of the maximum value at an endpoint of a domain. (The domain ends at $x = 0$ because $\sqrt{x}$ is not defined for negative values of x.)

4. THE SLOPE OF A TANGENT LINE

If we accept as true the assertion that a maximum can occur only under one of the three conditions above, the derivative becomes a marvelous tool for locating the value (or values) x for which the maximum of f occurs. We will now work through an example of the use of the derivative in locating the value x for which $f(x)$ is maximum. The example we give is a simple one, but it is not quite so simple that one can see the answer without the use of the derivative.

Suppose a farmer has 20 feet of wire fence that he wants to make into a rectangular pen, using his barn as one of the sides, as in Figure 12. What should the dimensions of the pen be if the area of the pen is to be as large as possible?

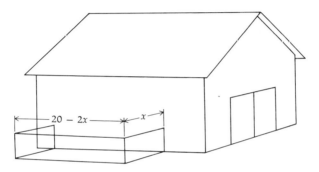

FIGURE 12. The problem is to maximize the area of a rectangular pen built up against the side of a barn. The total amount of fence used in the pen is 20 feet.

Letting x be the length of each of the two pen sides which will be perpendicular to the barn wall, we find that the area of the pen is the function

$$f(x) = (20 - 2x) \, x$$
$$= 20x - 2x^2.$$

From the geometry of the problem, we see that the largest value x can have is 10, and the smallest value x can have is 0. That is, the domain for the function representing the area of the pen consists of all numbers from 0 to 10. The numbers 0 and 10 are possibilities for the number x for which $f(x)$ is maximum, because they are the endpoints of the domain of f. However, we can dismiss them right away, because the area of the pen is 0 for each of those values.

No other values of x fall into the third category, so we now focus our attention on the derivative of f.

$$\frac{f(x+a)-f(x)}{a} = \frac{20(x+a)-2(x+a)^2-20x+2x^2}{a}$$

$$= \frac{20x+20a-2x^2-4xa-2a^2-20x+2x^2}{a}$$

$$= 20-4x-2a$$

which approaches $20-4x$ as a approaches 0.

Therefore the derivative exists for all values of x in the domain, so there are no values of x which fall into the second category. (Such is usually the case in problems of a geometrical or physical nature.) All we have left to do is find out those values of x for which $f(x)$ equals 0:

$$0 = 20-4x.$$

That is,

$$x = 5.$$

Therefore $x=5$ is the *only* candidate for a value of x at which $f(x)$ could be maximum. From the interpretation of the function f as the area of the pen, we are certain that there must be a maximum value, so we conclude that that maximum area must be

$$f(5) = 20 \cdot 5 - 2 \cdot 5^2$$
$$= 100 - 50$$
$$= 50 \text{ square feet.}$$

Notice that the rectangle which has the largest area is *not* a square. The rectangle of maximum area is 10 feet long and 5 feet wide. In contrast with the 50 square feet area of this rectangular pen, the area of a square pen made with 20 feet of fence and one barn wall is only $44\frac{4}{9}$ square feet.

You will notice that for quite some time now we have been talking only about the problem of finding the maximum value of a function. The primary reason for temporarily setting the minimum aside was to focus our attention better on the problem of finding the maximum. Actually, all that we have learned about finding the maximum value applies equally well to finding the minimum value. Any value x for which a function takes on its minimum value must also fall into one of the three categories we described before. For example, the minimum area of the rectangular pen, 0, occurs for both $x=0$ and $x=10$, the two endpoints of the domain of the area function.

One of the first mathematicians to develop a method for finding maxima and minima was Pierre de Fermat. His method was the same in spirit as the method we have studied in this section; had he continued his investigations, he may well have been led to the discovery of calculus. Fermat had a particular interest in maxima and minima, in connection with a conjecture he made which is now called the *principle of least time*. Fermat suggested that the path taken by a ray of light from one point to another will be such as to minimize the amount of *time* required for it to travel between the two points. His conjecture turned out to be correct, with one modification: Instead of the time always being minimized, rather the derivative of the travel time is always 0. (Usually, the time is a minimum as well.) Fermat's principle was the forerunner of the great *variational* principles in physics, which have proved to be of immense value in both relativity theory and quantum mechanics.

You probably have seen the trick illustrated in Figure 13, where the penny in the bottom of the cup, at first hidden by the rim, becomes visible when the glass is filled with water. If we adopt the viewpoint of Fermat's principle, the ray of light seems anxious to leave the water quickly, in order to get up into the air. Therefore, if it is minimizing its travel time, its speed in water must be slower than its speed in air. Using nothing more than the methods of this section, it is possible to calculate the ratio of the speed of light in air to the speed of light in water from a knowledge of the angle of bending. The speed ratio predicted in this manner (approximately 1.33) was verified two centuries after the time of Fermat, when it finally became possible to measure the speed of light.

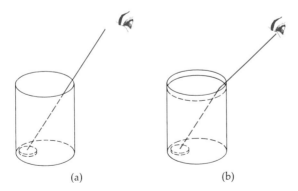

FIGURE 13. Illustration of a trick based on the bending of a light ray when it passes from water to air. (a) Without water, penny not visible. (b) With water, penny visible. According to Fermat's principle of least time, the direction of the bending tells us that light travels more slowly in water than it does in air.

EXERCISES

1. Find the minimum value of the function
$$f(x) = x^2 - x + 1.$$
What, then, is the range of this function? (This completes our information on Example 3 in Section 2.)

2. Find the maximum value of the function
$$f(x) = 32x - x^4.$$

3. Suppose the farmer mentioned in this section had chosen to build his pen *away* from the barn, thus using his 20 feet of fence for all four sides of the pen. Use the derivative technique to calculate the maximum area the pen can have under these circumstances.

4. A banking concern has found over the years that if it awards x percent yearly interest on deposits, then its total deposits will be x times 50 million dollars. It is able to loan out at an average $7\frac{1}{2}$ percent interest all the money it has on deposit. Write the bank's profit as a function $f(x)$. Use the derivative to determine what the value of x should be to maximize the bank's profit.

5. Repeat Exercise 4 for the case where the bank's total deposits will be x times 75 million dollars. Can you explain why the answers to Exercise 4 and this exercise are related the way they are?

6. Suppose you are in the business of making super-special bookends. The first pair of ends of a certain design will cost 100 dollars, and each succeeding pair will cost 5 dollars. You estimate that you can sell 600 pair, and that if you sell n pair, the selling price per pair will be $10 - n/100$ dollars. (The more you make, the less rare they are, so the cheaper the price.) How many pair should you make to maximize your profit?

7. The volume of a cylindrical tin can is equal to $\pi r^2 h$, where r is the radius of the base and h is the height. A can manufacturer wishes to produce a can of capacity 64 cubic inches. Therefore he must choose h so that
$$h = \frac{64}{\pi r^2}.$$
To make the can, he cuts the top and bottom out of two squares of sheet metal (the parts cut off being waste) and makes the side

from a rectangle (with no waste). What should be the radius of the base in order to minimize the amount of sheet metal used (including the waste)?

8. Repeat Exercise 7, except minimize the amount of metal of which the can is made. (Exercise 7 could be inspired by cost considerations, while this exercise could be inspired by weight considerations.)

9. Suppose a light source, a mirror, and someone's eyeball are located as in the figure below. Everything in the sketch is assumed to be in the air, so minimizing the travel time of light is the same as minimizing the distance of travel. Show that the distance traveled by a beam of light from the source and reflected by the mirror to the eyeball is minimized when the distance $x = 1$ foot.

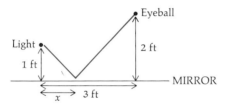

10. (For those familiar with computer programming.) Write a program to calculate the values of $32x - x^4$ for $x = .01, .02, \cdots, 3.00$. Run your program and determine which of the 300 values of x used gave the largest value of $32x - x^4$. Compare with the answer to Exercise 2.

11. (For those familiar with computer programming.) Write a program to search out the value of x to .00001 accuracy for which $32x - x^4$ is maximized.

5. Sine and Cosine

You may have been wondering about the need for the general notion of a function, inasmuch as all the functions we have studied so far have been defined merely by algebraic operations. We will now consider two functions that *cannot* be defined by a (finite) number of algebraic operations. And it happens that these two functions are tremendously important in applications of calculus. They are perhaps even more important than any of the functions we have seen so far.

The two functions are known as *sine* and *cosine*. A function is any rule that assigns to a number x a number $f(x)$. In the case of sine and cosine, the rules are completely geometric. Consider a circle of radius 1, drawn with its center at the origin of a pair of axes, as in Figure 14. Pick any number r, positive, negative, or 0. If r is positive, locate the point P which is r units along the circumference of the circle in a counterclockwise direction from the point marked START. In case r is larger than 2π, r units *along the circumference* entails traveling at least one full revolution around the circle and then some.

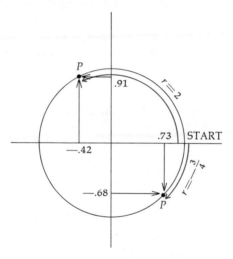

FIGURE 14. Illustration of the definition of sine and cosine. To two decimal places accuracy, $\sin 2 = .91$, $\cos 2 = -.42$, $\sin -\frac{3}{4} = -.68$, and $\cos -\frac{3}{4} = .73$.

If r is negative, locate the point P which is r units along the circumference of the circle in a clockwise direction from the point marked START. If $r = 0$, let P be the point marked START.

Having located the point P according to the above instructions, we define

$\sin r =$ the height of P above the horizontal axis (positive height being in the upward direction)

$\cos r =$ the distance of P from the vertical axis (positive distance being toward the right).

The most important feature to remember about sine and cosine is that the domain of each consists of the collection of *all* numbers. (Trigonometry courses frequently devote many weeks to solving triangle problems using sine and cosine, and as a result a person can begin to think of the domain of each as being rather restricted.) The ranges of sine and cosine each consist of all numbers from −1 to 1.

The second most important property of sine and cosine is their *periodicity*. As you can see from Figure 15, each function repeats itself every 2π units to the left and to the right. The reason for the periodicity lies in the definition. Traveling 2π units along the circumference of the circle in either direction brings one right back to the point from which he started.

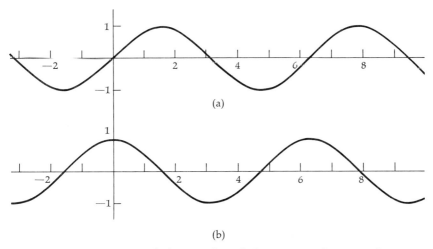

FIGURE 15. Portions of the graphs of the sine and cosine functions. (a) Sine. (b) Cosine. Each function has the collection of all numbers as its domain and the collection of all numbers from −1 to 1 as its range. Also, each function repeats itself every 2π units to the left and to the right.

Many different physical phenomena are periodic or are nearly periodic. Every swing of a pendulum, neglecting friction, is an exact repetition of the preceding swing. Some other examples of periodic (or nearly periodic) phenomena are: vibration of a guitar string, vibration of an ear drum, ocean waves, radio and television signals, electric house current, vibration of the nucleus of an atom.

One might think, then, that by studying the properties of sine and cosine it should be possible to get some small insight into periodic phenomena. Actually, much more is true. Not only are all of the phenomena listed periodic; every one of them can be described in terms of one or several sine functions. If you have a friend who is taking an electronics laboratory course, he can display ordinary house current on an oscilloscope for you. The curve you will see will be almost exactly that of Figure 15(a).

A person begins to wonder why there should be such unity in nature, that so many vibrations of so many different kinds should all be describable in terms of sine functions. The answer lies in calculus, in a remarkable property of the derivatives of sine and cosine. Unfortunately, we are not in a position to calculate the derivatives of sine and cosine. In order to calculate them, we would need a formula which is too difficult to be proved here. However, it is possible to examine the graphs of sine and cosine, draw a few tangent lines, and take an educated guess as to what the derivative of each is.

For example, let us take just one cycle of the sine function and draw the nine tangent lines shown in Figure 16. Actual measurement of the slopes of the tangent lines at $r = 0$ and $r = \frac{1}{4}\pi$ shows that the slopes are (as carefully as we can measure) 1 and .71. The tangent line at $r = \frac{1}{2}\pi$ has slope 0, because sine attains its maximum at that value. From the symmetry properties of the graph, we can conclude that the remaining six slopes are $-.71, -1, -.71, 0, .71,$ and 1.

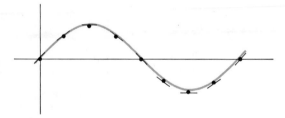

FIGURE 16. Nine tangents drawn on the graph of sin r, in an effort to see what the derivative of sin r might be.

In Figure 17, we have these nine values of the slopes plotted. They fall right into the pattern of the graph of the cosine function. In fact, the derivative of sin at the value r is always equal to cos r. And the derivative of cos r, in turn, is equal to $-\sin r$. The equations describing the various kinds of vibratory motion we listed above all

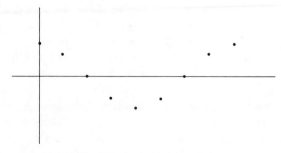

FIGURE 17. Values of the slopes of the nine tangent lines drawn in Figure 16. The points all lie on the graph of cos x.

5. SINE AND COSINE

involve derivatives. (They are called *differential equations*, somewhat inaccurately.) It just happens that the behavior of sine and cosine in trading places (with one minus sign thrown in) when we take their derivatives causes them to be solutions of those particular equations. And, as solutions to those equations, sine and cosine are among the most important functions in all of mathematics.

EXERCISES

1. For each of the following numbers x find (approximately if necessary) a number x' which is less than 2π but greater than or equal to 0 such that

$$\sin x' = \sin x \quad \text{and} \quad \cos x' = \cos x:$$

$$x = -\pi \qquad x = 3\pi \qquad x = 1$$
$$x = 10 \qquad x = -100.$$

2. Draw lines tangent to the graph of $\cos x$ for the five values $x = 0$, $\frac{1}{4}\pi, \frac{1}{2}\pi, \frac{3}{4}\pi$, and π. Measure the slopes of the lines as best you can, and plot these five slopes on a graph similar to the one in Figure 17. [*Hint:* Graph $\cos x$ as large as you can and as accurately as you can, or you will not have much success.]

3. Let f'' denote the function which is the derivative of the derivative of f. Verify that both $f(x) = \sin x$ and $f(x) = \cos x$ satisfy the equation

$$f'' = -f.$$

(The equation above is the equation of motion of a certain mass attached to the end of a spring.)

4. The motion of a pendulum is given (approximately) by the function

$$f(x) = c \sin (\sqrt{32/L}\, x),$$

where c is the maximum distance the pendulum is displaced from its rest position, where L is the distance from the mass of the pendulum to the pivot (both distances measured in feet), and where x is time elapsed, measured in seconds such that the pendulum passed through its rest position when $x = 0$. (See the figure that follows.) If a person is on a swing whose seat is tied 20 feet below a tree branch and is *not* pumping, how long does it take him to complete one full cycle of swinging? Answer the same question for a distance of 40 feet from the swing seat to the tree branch.

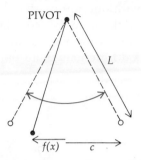

5. Pendulums were once commonly used in clocks as time standards. The usual length from mass to pivot was such that exactly 2 seconds were required for one complete swing. How long was the usual clock pendulum?

6. (For those familiar with computer programming.) The FORTRAN language contains built-in programs to calculate sine and cosine. To set the variable y equal to sin x, all you need to do is write the statement

$$Y = SIN(X)$$

or possibly

$$Y = SINF(X)$$

(You will need to use the second version only if your computer is of a fairly old vintage; check the computer manual if you are in doubt.) Write and run a program to calculate and print out SIN(X) for the values X = 0, .01, .02, $\cdots$, 7.00. For approximately what value of X between 3 and 4 is SIN(X) equal to 0? For approximately what value of X between 6 and 7 is SIN(X) equal to 0?

7. (For those familiar with computer programming.) For

$$f(x) = \sin x$$
$$x = 1, 2, 3, 4, \quad \text{and} \quad a = .01, .001, .0001, .00001,$$

write a program to calculate and print out the quotient (2) of Section 3. Run your program and compare the output with the values cos 1, cos 2, cos 3, and cos 4. [*Note:* Use a *radian* measure trigonometry table to look up the values of cosine. Or use one of the statements Y = COS(X), Y = COSF(X) to make the computer calculate the values for you.]

6. The Foundations of Calculus

With all the discoveries that Newton and Leibniz made in calculus, there was one aspect of the subject they left alone. That aspect, the logical foundations of the subject, was not studied adequately until nearly two centuries more had passed. The major task of the study of the foundations is to produce a workable definition of the word *approach* that we used in defining the derivative. For us, just beginning to explore the subject, it was reasonable to go on our intuition of the meaning of *approach*. However, as one goes deeper into calculus, it becomes important that the word be defined.

The earliest attempts toward developing the foundations were little more than double-talk. Leibniz had used the symbol

$$\frac{df}{dx}$$

for the number approached by the ratio

$$\frac{f(x') - f(x)}{x' - x}. \tag{1}$$

The df is a reminder that the numerator of (1) is a difference of two values in the range of f, and the dx is a reminder that the denominator of (1) is a difference of two values in the domain of f.

But then mathematicians began to talk about df and dx as numbers. The number df was taken to be the difference $f(x') - f(x)$, and the number dx was taken to be the difference $x' - x$. But x' was supposed to be chosen so that dx was "infinitely small," so that the ratio

$$\frac{df}{dx}$$

would be *exactly* equal to the slope of the tangent line to the graph. On the other hand, dx had to be different from 0, or else one could not divide by it. In everyday language, this is called "having your cake and eating it too."

The double-talk worked beautifully, however. By making the derivative into a fraction, it made it easy to think about. It thus helped along an intensely creative period of mathematics in which the new subject of calculus was developed and applied to an amazing number of different problems. Therefore no one did much complaining about the double-talk except the Irish philosopher-theologian Bishop George Berkeley.

He knew quite well that there was a logical difficulty in the number dx being infinitely small (that is, equal to 0) and yet not equal to 0 at the same time. He also was seeing the rise of a rationalistic philosophy which demanded that religion be held to the same standards of logical rigor that mathematics satisfied. So he did exactly what he was in position to do. With one stroke he knocked the shaky props right out from under calculus, saying as he did it that theology might not be logically sound, but neither was the greatest of all mathematical inventions, calculus. Berkeley was absolutely correct in his criticism of calculus, but mathematicians continued to believe what they wanted to believe about dx until the nineteenth century, when the concepts of *approach* and *limit* were defined without sleight-of-hand.

The definitions are complicated, and they are still regarded by some users of calculus as unnecessary. Many physics and engineering textbooks continue to use df and dx as "infinitesimal" differences, just as they were used three centuries ago. Because the definitions are so complicated, they create a dilemma in the teaching of first-year calculus. If they are covered first, the student's attention promptly wanders, because he does not know how necessary they are to the definition of the derivative. On the other hand, if the instructor begins with the derivative, it seldom happens that he will go back to the preceding definitions, so the structure of calculus built up in the student's mind is a bit on the shaky side.

Since we were led into the discussion of foundations by our intuitive idea of *approach*, let us conclude by displaying an adequate definition of the word. Those of you who will study calculus no further can examine it for its complexity and agree that no wonder it was a while before the foundations of calculus were developed. Those of you who will study calculus further can do the same as well as keep the definition tucked away for reference later on.

Let g be a function with domain D. Let x be a number such that for every number δ greater than 0 there are infinitely many numbers x' in D whose distances from x are less than δ.

DEFINITION. The function g is said to *approach* the number b as x' approaches x if:

for every number ϵ greater than 0,
there exists a number δ greater than 0,
such that when x' is in D and x' is less than δ units from x,
then $g(x')$ and b are less than ϵ units apart.

6. THE FOUNDATIONS OF CALCULUS

Informally, the definition is saying that if we want to make $g(x')$ and b close together, we can do it by taking any x' which is sufficiently close to x. But it takes all those words to say it precisely!

EXERCISE

1. We used the symbol g rather than f in the definition of *approach* in order that we could more easily combine this definition with the definition of the derivative. To wit, we can let

$$g(x') = \frac{f(x') - f(x)}{x' - x}.$$

Try combining the two definitions, so as to obtain a definition of the derivative which does not use the word *approach*. (The result is a precise but very complicated definition of the word *derivative*.)

Appendix

TABLE A

Probabilities of obtaining (under the null hypothesis) a result at least as extreme as the one obtained, tabled for values of x ranging from 0 to 9.9.

x	\multicolumn{10}{c}{Next Decimal Place in x}

x	0	1	2	3	4	5	6	7	8	9
.0	1	.9203	.8875	.8625	.8415	.8231	.8065	.7913	.7773	.7642
.1	.7518	.7401	.7290	.7184	.7083	.6985	.6892	.6801	.6714	.6629
.2	.6547	.6468	.6390	.6315	.6242	.6171	.6101	.6033	.5967	.5902
.3	.5839	.5777	.5716	.5657	.5598	.5541	.5485	.5430	.5376	.5323
.4	.5271	.5220	.5169	.5120	.5071	.5023	.4976	.4930	.4884	.4839
.5	.4795	.4751	.4708	.4666	.4624	.4583	.4543	.4503	.4463	.4424
.6	.4386	.4348	.4310	.4274	.4237	.4201	.4166	.4131	.4096	.4062
.7	.4078	.3994	.3961	.3929	.3897	.3864	.3833	.3802	.3771	.3741
.8	.3711	.3681	.3652	.3623	.3594	.3566	.3537	.3510	.3484	.3455
.9	.3428	.3401	.3375	.3349	.3323	.3297	.3272	.3247	.3222	.3197
1.0	.3173	.3149	.3125	.3102	.3078	.3055	.3032	.3009	.2987	.2965
1.1	.2943	.2921	.2899	.2878	.2857	.2835	.2815	.2794	.2774	.2753
1.2	.2733	.2713	.2694	.2674	.2655	.2636	.2617	.2598	.2579	.2560
1.3	.2542	.2524	.2506	.2488	.2470	.2453	.2435	.2418	.2401	.2384
1.4	.2367	.2351	.2334	.2318	.2301	.2285	.2269	.2253	.2238	.2222
1.5	.2207	.2191	.2176	.2161	.2146	.2131	.2117	.2102	.2088	.2073
1.6	.2059	.2045	.2031	.2017	.2003	.1990	.1976	.1963	.1949	.1936
1.7	.1923	.1910	.1897	.1884	.1871	.1859	.1846	.1834	.1821	.1809
1.8	.1797	.1785	.1773	.1761	.1750	.1738	.1726	.1715	.1703	.1692
1.9	.1681	.1670	.1659	.1648	.1637	.1626	.1615	.1604	.1594	.1583
2.0	.1573	.1563	.1552	.1542	.1532	.1522	.1512	.1502	.1492	.1483
2.1	.1473	.1463	.1454	.1444	.1435	.1426	.1416	.1407	.1398	.1389
2.2	.1380	.1371	.1362	.1354	.1345	.1336	.1328	.1319	.1311	.1302
2.3	.1294	.1285	.1277	.1269	.1261	.1253	.1245	.1237	.1229	.1221
2.4	.1213	.1206	.1198	.1190	.1183	.1175	.1168	.1160	.1153	.1146
2.5	.1138	.1131	.1124	.1117	.1110	.1103	.1096	.1089	.1082	.1075
2.6	.1069	.1062	.1055	.1049	.1042	.1036	.1029	.1023	.1016	.1010
2.7	.1002	.0997	.0991	.0985	.0979	.0973	.0966	.0960	.0954	.0949
2.8	.0943	.0937	.0931	.0925	.0919	.0914	.0908	.0902	.0897	.0891
2.9	.0886	.0880	.0875	.0869	.0864	.0859	.0853	.0848	.0843	.0838

TABLE A *(continued)*

x	\multicolumn{10}{c}{Next Decimal Place in x}									
	0	1	2	3	4	5	6	7	8	9
3.0	.0833	.0828	.0822	.0817	.0812	.0807	.0802	.0797	.0793	.0788
3.1	.0783	.0778	.0773	.0769	.0764	.0759	.0755	.0750	.0745	.0741
3.2	.0736	.0732	.0727	.0723	.0719	.0714	.0710	.0706	.0701	.0697
3.3	.0693	.0689	.0684	.0680	.0676	.0672	.0668	.0664	.0660	.0656
3.4	.0652	.0648	.0644	.0640	.0636	.0633	.0629	.0625	.0621	.0617
3.5	.0614	.0610	.0606	.0603	.0599	.0595	.0592	.0588	.0585	.0581
3.6	.0578	.0574	.0571	.0567	.0564	.0561	.0557	.0554	.0551	.0547
3.7	.0544	.0541	.0538	.0534	.0531	.0528	.0525	.0522	.0519	.0516
3.8	.0513	.0509	.0506	.0503	.0500	.0497	.0495	.0492	.0489	.0486
3.9	.0483	.0480	.0477	.0474	.0472	.0469	.0466	.0463	.0460	.0458
4.	.0455	.0429	.0404	.0381	.0359	.0339	.0320	.0302	.0285	.0269
5.	.0253	.0239	.0226	.0213	.0201	.0190	.0180	.0170	.0160	.0151
6.	.0143	.0135	.0128	.0121	.0114	.0108	.0102	.0096	.0091	.0086
7.	.0082	.0077	.0073	.0069	.0065	.0062	.0058	.0055	.0052	.0049
8.	.0047	.0044	.0042	.0040	.0038	.0036	.0034	.0032	.0030	.0029
9.	.0027	.0026	.0024	.0023	.0022	.0021	.0019	.0018	.0017	.0017

TABLE B

Critical values of $|V|$ in the Wilcoxon test.

m \ n	1	2	3	4	5	6	7	8	9	10
1	...	...	...	...	...	...	...	...	...	...
2	...	...	...	...	...	...	...	8.0	9.0	10.0
3	...	...	...	...	7.5	8.0	9.5	10.0	11.5	12.0
4	...	...	...	8.0	9.0	10.0	11.0	12.0	14.0	15.0
5	...	...	7.5	9.0	10.5	12.0	12.5	14.0	15.5	17.0
6	...	...	8.0	10.0	12.0	13.0	15.0	16.0	17.0	19.0
7	...	...	9.5	11.0	12.5	15.0	16.5	18.0	19.5	21.0
8	...	8.0	10.0	12.0	14.0	16.0	18.0	19.0	21.0	23.0
9	...	9.0	11.5	14.0	15.5	17.0	19.5	21.0	23.5	25.0
10	...	10.0	12.0	15.0	17.0	19.0	21.0	23.0	25.0	27.0
11	...	11.0	13.5	16.0	18.5	20.0	22.5	25.0	26.5	29.0
12	...	11.0	14.0	17.0	19.0	22.0	24.0	26.0	28.0	31.0
13	...	12.0	15.5	18.0	20.5	23.0	25.5	28.0	30.5	32.0
14	...	13.0	16.0	19.0	22.0	25.0	27.0	30.0	32.0	34.0
15	...	14.0	17.5	20.0	23.5	26.0	28.5	31.0	33.5	36.0
16	...	15.0	18.0	21.0	25.0	27.0	30.0	33.0	35.0	38.0
17	...	15.0	19.5	23.0	25.5	29.0	31.5	34.0	37.5	40.0
18	...	16.0	20.0	24.0	27.0	30.0	33.0	36.0	39.0	42.0
19	...	17.0	21.5	25.0	28.5	32.0	34.5	38.0	40.5	43.0
20	...	18.0	22.0	26.0	30.0	33.0	36.0	39.0	42.0	45.0

TABLE B (continued)

Critical values of $|V|$ in the Wilcoxon test.

m \ n	11	12	13	14	15	16	17	18	19	20
1	...	...	...	...	...	...	...	...	...	...
2	11.0	11.0	12.0	13.0	14.0	15.0	15.0	16.0	17.0	18.0
3	13.5	14.0	15.5	16.0	17.5	18.0	19.5	20.0	21.5	22.0
4	16.0	17.0	18.0	19.0	20.0	21.0	23.0	24.0	25.0	26.0
5	18.5	19.0	20.5	22.0	23.5	25.0	25.5	27.0	28.5	30.0
6	20.0	22.0	23.0	25.0	26.0	27.0	29.0	30.0	32.0	33.0
7	22.5	24.0	25.5	27.0	28.5	30.0	31.5	33.0	34.5	36.0
8	25.0	26.0	28.0	30.0	31.0	33.0	34.0	36.0	38.0	39.0
9	26.5	28.0	30.5	32.0	33.5	35.0	37.5	39.0	40.5	42.0
10	29.0	31.0	32.0	34.0	36.0	38.0	40.0	42.0	43.0	45.0
11	30.5	33.0	34.5	37.0	38.5	41.0	42.5	44.0	46.5	48.0
12	33.0	35.0	37.0	39.0	41.0	43.0	45.0	47.0	49.0	51.0
13	34.5	37.0	39.5	41.0	43.5	45.0	47.5	50.0	51.5	54.0
14	37.0	39.0	41.0	43.0	46.0	48.0	50.0	52.0	55.0	57.0
15	38.5	41.0	43.5	46.0	48.5	50.0	52.5	55.0	57.5	60.0
16	41.0	43.0	45.0	48.0	50.0	53.0	55.0	58.0	60.0	62.0
17	42.5	45.0	47.5	50.0	52.5	55.0	57.5	60.0	62.5	65.0
18	44.0	47.0	50.0	52.0	55.0	58.0	60.0	63.0	65.0	68.0
19	46.5	49.0	51.5	55.0	57.5	60.0	62.5	65.0	67.5	71.0
20	48.0	51.0	54.0	57.0	60.0	62.0	65.0	68.0	71.0	73.0

TABLE C

Critical values of $|S|$ in the Kendall test

Sample Size	Critical Value	Sample Size	Critical Value
1	...	21	66
2	...	22	71
3	...	23	75
4	8	24	80
5	10	25	86
6	13	26	91
7	15	27	95
8	18	28	100
9	20	29	106
10	23	30	111
11	27	31	117
12	30	32	122
13	34	33	128
14	37	34	133
15	41	35	139
16	46	36	146
17	50	37	152
18	53	38	157
19	57	39	163
20	62	40	170

Index

Abacus, 271
Absorbing Markov matrix, 192
Abundant number, 16
ACA and ADA, 235
Addition of matrices, 154
  associative law for, 156
  commutative law for, 156
Adjacency matrix, 159
Alcoholism and religion, 229–232
Algebra (corruption of *al-jabr*), 38
Algorithm, 37–38
  Euclid's, 32–41
ALGOL, 248, 272
Alice, 251
Al-Khowârizmi, Abu, 38
Anniversaries, 111–114
Approach, 306
  used to define derivative, 287
Arabic number system, 128
Aristotle, 285
Assignment statement, 252–253
  not an algebraic equation, 257

Associative law(s), 156, 161, 167, 174
Authoritarianism, 241–242
Axis of a graph, 280

Babbage, Charles, 271
Backus, John, 248
Barrow, Isaac, 289
BASIC, 272, 274
Bending of light, 297
Berkeley, George, 305–306
Binary number system, 128–132
  conversion to and from decimal, 130–131
  in liquid measure, 132
  in Nim, 133–135
Binomial coefficients, 197–201
Binomial theorem, 202
Birth rate, monthly variation of, 121
Birthday, Washington's, 113
Birthdays, matching, 118–122
  and sex, 222–223
Blackjack, 273

318 INDEX

Bouton, C. L., 133
Brahma, Tower of, 137–142
Braid(s), 116–118
 split-strap, 117–118
 standard, 116
Branching (IF) statement, 256
Bridge problem, Königsberg, 63–66

Calculus, 277–307
Cancer, lung, 217–218
Card programming, 274–276
Cardano, Girolamo, 193
Cartesian coordinates, 280–281
Casting out elevens, 126
Casting out nines, 122–127
 errors detectable by, 126
Cayley, Arthur, 93
CEEB, 235
Central limit theorem, 202
Check column, 182
 use of, 186–187
Checkerboard, domino covering of, 7–10
 in study of Fifteen Puzzle, 144
 tromino covering of, 10
Chess, 142
Cockcroft, John, 1
Coding paper, FORTRAN, 276
Coin, fair, 195
 flipping, 195–209
 imbalance in pennies, 210–211
Commas in FORTRAN, 249–250, 275
Commutative law, 156
 not satisfied by matrix multiplication, 156–157
Composite number(s), 18
 existence of arbitrarily long strings of, 24
 infinitude of, 19
Computer(s), 247–276
 and binary number system, 129
 generations of, 272
 programs as algorithms, 38
Conditional transfer (IF) statement, 256
Connected graph, 64
Constant, FORTRAN, 249–250, 275

Constraint edge, 72–73
Contingency table, 218
Contraception, 169
Correlation, 220, 237
Cosine, 300–304
Creativity in mathematics, periods of, 3–4
Critical path, 76
 and slack times, 78
Critical value of test statistic, 213
Cube, 85–90
Cyclic Markov matrix, 192

Dating and parental income, 237–240
Decimal, unending, 46
Decimal number system, 124–125, 128–129
 conversion to and from binary, 130–131
Decimal point, 128
 in FORTRAN constant, 250
Deficient number, 16
De Moivre, Abraham, 202
De Morgan, Augustus, 92–93
Derivative(s), 287–307
 method for computing, 289–290
 of sine and cosine, 302–303
Descartes, René, 61
 formula for graphs, 79
 formula for polyhedra, 78
 graphing technique for functions, 280
Diagonal entries of matrix, 161
Differential equations, 303
Digit, 128–129
Digital computer (*see* Computer)
Distributive laws, 155
Dodecahedron, regular, 85–90
Domain of a function, 278
Dominoes, analogy to mathematical induction, 141
 covering checkerboard, 7–10
Duality, 67–70
 among polyhedra, 90

Edge, constraint, 72–73
 of graph, 64, 158

INDEX 319

of polyhedron, 78
task, 72
Eighth bridge of Königsberg, 66
Elementary row operations, 182–183
Elements, association with regular polyhedra, 85–86
Elevens, casting out, 126
ENIAC computer, 271–272
END statement, 253
Eratosthenes, 21
Escher, M. C., 62
Euclid, 15–16
  algorithm, 32–41
  Elements, 15–16, 33
Euler, Leonhard, 25, 61
  factorization of $2^{25} + 1$, 27
  formula for graphs, 79
  formula for polyhedra, 78
  solution of Königsberg bridge problem, 64–66
  theorem on graph tracing, 65
Even rearrangement, 145
Extra ordinary map, 101–105

Face of polyhedron, 78
Factorial, 20, 199
Factorization, of $x^b - 1$, 30
  of $x^b + 1$, 26
  of $2^{25} + 1$, 27
Fair coin, 195
Family closeness, study of, 218–222
Fermat, Pierre de, 27, 193, 289
  conjecture on Fermat numbers, 27
  last problem, 58–59
  principle of least time, 297
Fifteen Puzzle, 142–150
Fingers, used for counting, 129
Flipping coins, 195–209
Flowcharting, 261–264
Fluxion, 289
FORMAT statement, 253
FORTRAN, 247–276
FORTRAN formula, 248–251
FORTRAN statement(s), assignment, 252–253
  conditional transfer (IF), 256
  END, 253

  FORMAT, 253
  output (PRINT), 253
  STOP, 253
  unconditional transfer (GO TO), 268
Foundations, of calculus, 305–307
  of probability theory, 194
Fraternities, 232–233
Frequency interpretation of probability, 194
Function, 278

Galilei, Galileo, 285, 289
Garfield, James A., 47
Gematriya, 17
Gillies, Donald, 29
Goldbach, Christian, 23
Graph(s), 64, 158
  adjacency matrix of, 159
  applied to business management, 70–78
  applied to map coloring, 94–100
  Descartes-Euler formula for, 79
  of functions, 280–283
  PERT charts, 70–78
  as road maps, 158–159
Greatest common divisor, 32–41
  as a linear combination, 34
Gregory XIII, Pope, 113
Growth of a population, Malthus' model, 168–169
  matrix model, 164–171
Guthrie, Francis, 92–93

Hamiltonian circuit, 67
Handshaking, mathematics of, 67
Heawood, P. J., 93
  five-color theorem for plane maps, 93
  seven-color theorem for doughnut maps, 108–109
High-level language, 247–248
Highway inspector, 66
Hindu number system, 38, 128
How to Lie with Statistics, 216
Hypothesis, null, 208, 216
  rejection of, 207–208

Icosahedron, regular, 85–90
IF statement, 256
Induction, mathematical, 141
Infinitesimal, 306
Infinitude, of composite numbers, 19
  of prime numbers, 22

Jews, fraternity membership, 232
  low alcoholism rate, 229–232
Julius Caesar, 113

k (in Kendall test), 239
Kempe, A. B., 93
Kendall test, 235–240
  critical values of $|S|$, 314
Knots, 114–116
Königsberg bridge problem, 63–66
Kummer, E. 58–59

Last Year at Marienbad, 133
Lead 1, 184
Leap years, rule for determining, 112
Lehrer, T. A., 129
Leibniz, G. W., desk calculator, 271
  discovery of the derivative, 289
Length of path in a graph, 160
Light, bending of, 297
Linear combination, 33
Linear equations, solution, 180–187
  system, 179
List arithmetic, 152–154
Loops in FORTRAN, 256–264
  nested, 259–264
Loyd, Sam, 142, 147
Lucas, Edouard, 31, 137
Lung cancer, 217–218

Malthus, Thomas, 168–169
Mann-Whitney test, 226
Map(s), coloring problems, 92–110
  containing ring countries, 106–107
  on doughnuts, 107–109
  extra ordinary, 101–105
  four-color problem, 109
  ordinary, 98–100
  proof of five-color theorem, 102–105
  on spherical surface, 107
  viewed as graphs, 94–98
Mark I computer, 271–272
Markov, A. A., 171
Markov chain, 171–178, 188–192
  stability of, 175
Markov matrix, 171
  absorbing, 192
  cyclic, 192
Mathematical induction, 141
Mathematical statistics, 193
Matrix, 152–192
  arithmetic, 154
  check-column, 182
  diagonal entries of, 161
  Markov, 171
  row-reduced, 185
  transition, 166
Maximum of a function, conditions for, 294
Mersenne, Marin, 30
Mersenne numbers, 30–32
Minimum of a function, conditions for 296
Minimum time to completion in PERT, 74–76
Misuses of statistics, 215–216
Möbius, A. F., 62
Motion, of falling object, 285–288
  of mass on spring, 303
  mathematical study of, 285
  of pendulum, 303
Multiplication of matrices, 154–155
  associative law, 156, 161, 167, 174
Multiplication symbol, required in FORTRAN, 251

Natural numbers, 15–59
  abundant numbers, 16
  composite numbers, 18–21
  deficient numbers, 16
  gematriya, 17
  perfect numbers, 16, 31, 131
  prime numbers, 18–32
  supposed magic powers of, 16–17
Negative correlation, 220, 237
Nested loops in FORTRAN, 259–264

Neugebauer, Otto, 44
Newton, Isaac, 217
  discovery of the derivative, 289
Nim, 132–137
  rules, 133
  strategy, 133–134
Nines, casting out, 122–127
  errors detectable by casting out, 126
Null hypothesis, 208, 216
  rejection of, 207–208
Number (real number), 278
Number system(s), Arabic, 128
  base twenty, 129
  binary, 128–132
  conversion between binary and decimal, 130–131
  decimal, 124–125, 128
  Hindu, 128
  sexagesimal, 46

Octahedron, regular, 85–90
October revolution, 113
Odd rearrangement, 145
Odd vertex, 65
Ordinary map, 98–100
Origin of a graph, 280
Output (PRINT) statement, 253

p (in Pearson test), 219
Parental income and dating, 237–240
Parentheses, elimination by associative law, 156
  in FORTRAN, 249
  used to enclose matrices, 154
Pascal, Blaise, 193
  adding machine, 271
Path, critical, 76
  length of, in a graph, 160
Pearson, Karl, 219
  test, 217–224
Pendulum, 301, 303
Pennies, fairness of, 210–211
Perfect numbers, 16, 31, 131
Periodicity of sine and cosine, 301
PERT, 70–78, 163
Platonic solids, 85

Plimpton 322 (Babylonian tablet), 44–48, 57
PL/I, 248, 272
Polaris submarine project, use of PERT, 72, 76
Polyhedron, 78–91
  Descartes-Euler formula, 78
  regular, 85–91
Polynomial, 25
Population growth, Malthus' model, 168–169
  matrix model, 164–171
Positive correlation, 220, 237
President(s) of U.S., deaths in office, 17
  Garfield, James A., 47
  matching birthdays of, 122
  matching death days of, 122
  Washington, George, 113
Prime number(s), 18
  formula for, 28
  infinitude of, 22
  largest known prime, 29–31
  twin primes, 23
PRINT statement, 253
Probability, frequency interpretation, 194
  of obtaining r heads, 200
Probability theory, 193–194
Programmable calculator, 273–274
Programming languages, ALGOL, 248, 272
  BASIC, 272, 274
  FORTRAN, 247–276
  PL/I, 248, 272
Pythagoras, 11, 42, 85–86
Pythagorean Brotherhood, 11, 85–86
Pythagorean theorem, 42, 46–47
Pythagorean triples, 42–58
  formulas for, 48–49
  primitive triples, 44

Railroad, first U.S. transcontinental, 113
Randomness assumption, 221
Range of a function, 278
Rate your mind pal (puzzle), 147–148

Reaction times, 209–210
Rearrangements, 145
  proof of difference between even and odd rearrangements, 150–151
Refraction of light, 297
Regular polyhedra, duality among, 90
  theorem on, 86–87
Ringel, Gerhard, 109
Row operations, elementary, 182–183
Row-reduced form, 185
Row reduction process, 181–188

S (in Kendall test), 236–238
  critical values of $|S|$, 314
Sachs, A. J., 44
Salesman, automobile, 235
  Fly-by-Night, 66
  traveling, 162
Sample size, 216–217, 222, 243–246
Search procedure, program of a, 266
Seven as lucky number, 16
Sex ratio, human, 222–223
Sexagesimal number system, 46
Sieve of Eratosthenes, 21
Sign test, 209–210
Significance level, 207
  time to choose, 208
Simple polyhedron, 78
Simulation on a computer, 272–273
Sine, 300–304
Slope, 292
Smoking and lung cancer, 217–218
Speed, 285–288
  at a particular time, 287
Spirograph, 41
Square root of two, 10–12
Statement, FORTRAN (see FORTRAN statement)
Statistical tests, Kendall test, 235–240
  Pearson test, 217–224
  role in research, 216–217
  sign test, 209–210
  Wilcoxon test, 225–232
STOP statement, 253
  numbered, 257–258

Table A, 243, 310–311

Tangent line, 292
Task, slack time of, 77–78
Task edge, 72
Test statistic(s), 213
  k (Kendall), 239
  p (Pearson), 219
  random fluctuations of, 245
  $|S|$ (Kendall), 237
  $|V|$ (Wilcoxon), 228
  w (Wilcoxon), 231
  x, 202
Tetrahedron, 78, 85–90
Thinking and computers, 273
Thorp, Edward, 273
Time-sharing on computers, 274
Time to completion, minimum, in PERT, 74–76
Topological properties, examples, 61–62
Tower of Brahma, 137–142
Transfer statement, conditional (IF), 256
  unconditional (GO TO), 268
Transition matrix, 166
Transposition, 145
Tree diagram, 196
Trigonometry, and Plimpton 322, 44–46
  sine and cosine functions, 299–303
Tromino, 10
Twin primes, 23
Two, irrationality of square root, 10–12

U (in Wilcoxon test), 227, 230
Unconditional transfer (GO TO) statement, 268
Understanding mathematics, use of examples in, 2, 6–7
Unending decimal, 46
Updating PERT chart, 76
Utilities problem, 84–85

V (in Wilcoxon test), 227–231
  critical values of $|V|$, 312–313
Variable, FORTRAN, 250
Vertex of graph, 64, 158

Vertex of polyhedron, 78

w (in Wilcoxon test), 231
Wallis, John, 289
Walton, E. T. S., 1
Washington, George, 113
Wilcoxon, Frank, 226
Wilcoxon test, 225–232

critical values of $|V|$, 312–313
Wolfskehl, Paul, 59

x, 202
    as measure of extremeness, 203
    table of probabilities, 310–311

Youngs, J. W. T., 109